SHRINK TO GROW

Shrink to Grow

Lessons from Innovation and Productivity in the Electronics Industry

Jürgen Kluge
Ingo Beyer
Rupert Deger
Dieter Düsedau
Wolfgang Huhn
Edward Krubasik
Eberhard Schmidt
Lothar Stein

MACMILLAN
Business

English translation first published in the United Kingdom by
MACMILLAN PRESS LTD
Houndmills, Basingstoke, Hampshire RG21 6XS
and London
Companies and representatives
throughout the world

ISBN 0–333–67478–2

A catalogue record for this book is available from the British Library.

10 9 8 7 6 5 4 3 2
05 04 03 02 01 00 99 98 97 96

Copy-edited and typeset by Povey–Edmondson
Okehampton and Rochdale, England

Printed and bound in Great Britain by
Antony Rowe Ltd
Chippenham, Wiltshire

Contents

List of Figures and Tables

FIGURES

TABLES

Acknowledgements

This book builds on the breadth and level of detail of a database of the worldwide electronics industry compiled over recent years under the title 'Excellence in Electronics'. This database was only made possible through close cooperation between practice and science – between McKinsey & Company in Europe, the United States and Japan, and the University of Augsburg, Stanford University and Stanford Japan Center.

We extend our particular thanks to the many senior executives who participated in the long-term study on behalf of their companies and who took the time to answer the many detailed questions in the survey. Our thanks are also due to the University of Augsburg; without the commitment of Professor Coenberg and the tireless involvement of Dr Günther, Mrs Wingert and Mr Prillmann it would have been impossible to build and validate the European section of the database. In the United States we thank Stanford University and Professors Miller, Okimoto and Mendelsohn, who collected the North American data with the help of Mrs Price and Mr Tabrizi. We also thank Professor Imai and the Stanford Japan Center for the data gathering carried out in Japan by Mrs Yamazaki, Professor Seo, Mrs Kopczak, Mr Johnston, Mr Sullivan and Mr Fujihara. We also wish to thank the Ministry of International Trade and Industry (MITI), which established contacts with companies in Japan and lent its support to the project.

We wish to acknowledge especially the contributions of Johannes Ziegler and Art Cimento as leaders of the McKinsey project team; of Hans Gersbach and Koji Sakate of the McKinsey Global Institute in Washington DC; and of their partners at the University of Groningen, Messrs van Ark and Pilat. Validating the data and presenting the individual results to participating companies would have been a far more difficult task without the support of Ingo Wuggetzer, Andreas Bährle, and Christof Stierlen.

Substantial work interpreting and preparing the data was carried out in a number of workshops with members of the McKinsey Electronics Sector practice. In this respect, we are particularly grateful to our McKinsey colleagues Mike Nevens, Hubert Joly, Eric Labaye, Andreas Biagosch, Mahesh Krishnamurthy, Inder Sidhu and

Keiko Kin. In addition, in many places numerous individual observations made during conversations with senior company executives found their way into the study findings.

Without the indefatigable efforts of Gabriele Stimpfle there would never have been a clean manuscript on the exhausting road from database to book. And without the invaluable editorial guidance of Dagmar Boss and Christel Delker the manuscript would never have become ripe for publication. Our thanks are also due to Partha Bose, Helen Robertson and Robert Whiting for their work in preparing the English edition.

To all those named above, as well as to the many others who contributed in various ways to this book, we extend our heartfelt thanks.

Preface

To many observers and management gurus, electronics is one of the most absorbing industries there is. The industry's complexity and dynamism make it an outstanding experimental arena for sophisticated management: quantum leaps in the value and cost of the product offering, rapid change in product and manufacturing technology, investments in innovation so enormous that a company's very survival depends on getting it right. Because of these risks and opportunities, the art of management is thrusting forward more vigorously in this industry than in any other.

Reason enough for the constant recurrence of the electronics industry in 'excellent company' examples, from McKinsey and other sources: the best companies here act as role models beyond the industry. Reason enough, too, for our long-term study 'Excellence in Electronics', on which this book is based.

But a second major reason is the vital macroeconomic importance of the subject: every major industrialized nation in the world looks for a fair (or even more than fair) share in high-tech industries – with an accompanying distribution of inward investment. The result is a fierce battle between the leaders in the Triad of Europe, the USA and Japan for these industries, a battle intensified in the last decade by the appearance of new competitors such as the four Tiger States in the Far East, not to mention China.

For Europe this game has not gone very well so far. The high technology-based share of manufacturing is decreasing, particularly in electronics. Competition from other Triad regions seems to be gaining the upper hand. And this competition is increasingly entering domestic European markets, where home suppliers traditionally had the edge: closed regional markets offered certain industries protection in some countries from the full force of international competition – and some industries were protected throughout Europe. Those days are done for the computer industry and for consumer electronics, as well as for private and public telecommunications. Industrial electronics and measurement equipment will soon be exposed.

At McKinsey we know what is at stake from a number of consulting projects undertaken over the past few years to help construct defence strategies for European electronics, in which we helped various businesses to regain competitiveness or get a fresh start. European players, too, are becoming more aggressive. They are trying to create world-class structures on their own home turf and, as they do so, scale advantages are increasing in importance – presence in only one country is not enough. In addition, Europeans are increasingly seeking the tough challenge of the lead markets, because learning to fight in these hotly contested areas is the only way to receive the control signals that will keep a company in the world class.

And thus the Europeans in our firm became all the more interested in finding out, in a worldwide comparison, exactly what it took to achieve excellence in the electronics industry. But our colleagues in the US and Japan also perceived that every advantage in this Triad-wide battle was short-lived – it was, they thought, vital for all participants to understand the drivers of excellence. Finally, this was confirmed by the wide interest shown within the three regions observed, Europe, US and Asia. The curiosity of more than a hundred companies to track down the true success factors – to focus down on what drives or hampers return and growth – was soon aroused. As was their readiness to take an active part in the survey.

The findings are particularly hard on Europe in many areas, and the need for action is most urgent here. The results indicate clear levers for the management agenda of the next five years in finding the way back to world class, and in making the necessary changes in the corporate environment.

One message that came across particularly clearly was that the best companies pursue a policy of aggressive, continuous self-renewal. Those companies with the highest annual restructuring *and* growth rates are also the most successful. Both elements are essential, the one makes the other possible – and in both dimensions the measure of a successful company is much higher than any of us thought. The 'Excellence in Electronics' study has convinced us that it is not the *static difference* in cost or productivity that is decisive, but the *rate of improvement*, whether in cost reduction or innovation.

From many successful Total Redesign to World Class programmes, we have concluded that Europe should not throw in the towel, even in high-volume markets, and that in Germany, too, all electronics businesses can be made competitive. World class can be attained here too, but at a price. To attain world class, companies must be prepared to work very hard, never stand still, permanently challenge their processes and products, restructure every year and grow aggressively. Why that is so, and how it can be done, is described in the following chapters of this book.

If, then, the true art lies in self-renewal, which is an art that European managers will certainly not deny, then it follows that the frequently heard call for state support will be affected. Such support should not be invoked to protect individual industries or technologies, or give them special care and attention. But it should be invoked to create an environment which allows – or even forces – companies to renew themselves constantly: uncompromising competition, no limits to innovation, mobility of labour, continuous personal renewal of every individual by the best, continuous training.

A retreat into old, protectionist structures is practically inconceivable today in the international network of economies. A European solution with its own market laws requires more imagination than most of us have evinced up to now. And what is happening in electronics is clearly only an aggressive model for a whole range of other industries, such as automotive, mechanical engineering, pharmaceuticals, chemicals. The attempt to put our industry into training for world-class competition is an undertaking of macroeconomic dimensions.

How can we create enough growth to compensate for the restructuring effects of the unremitting race to close the gap? If current volumes in the electronics industry in Europe, which creates value-added requiring one million jobs, will be possible with only 400 000 in ten years' time, where is the growth and innovation to come from for 600 000 new jobs? A first estimate shows that 40 per cent of them will have to come from completely new businesses.

For those of us who have dedicated many years to the subject of innovation and technology management, the management tasks for the forthcoming transformation are an exciting prospect. It is no coincidence that it was mainly the technology experts from our firm who took a hand in the 'Excellence in Electronics' study. But because the results are of interest not only to them, and also not only to the electronics industry, the main messages have been summarized in this book. It is intended to spur on European companies to join the world-class players.

Munich EDWARD KRUBASIK

1 Introduction: A New World-Class Paradigm

In electronics, it is not the laggard that sets the pace. This industry, with its unique history of innovation and growth, passes on its way regardless of who goes to the wall. The failure of traditional approaches is hard on the weak players – neither rigorous restructuring nor aggressive expansion alone are enough to close the gap. And the fact that European companies hardly feature among the top players any more is hard on Europe. But there are grounds for optimism: no company holds life membership in the world-class league. There is still room at the top – but only for companies that play by the world-class game-rules: restructure and expand simultaneously, or 'trim to grow'.

The successes of the electronics industry are plain to see. Manufacturers renew their product lines at an astounding rate. Innovations appear in quick succession. Product generations are obsolete almost as soon as they hit the market. And the macroeconomic significance of this young industry has grown just as quickly. By the end of the decade, electronics, as a supporting technology used in manufacturing, office and the home, will account for 10 per cent of gross national product. Thirty per cent of all industrial production will contain a healthy portion of electronics, and about 90 per cent of all manufacturing processes will depend on electronic components.

But for the players in this global success story, developments are usually less predictable. What looks like a continuous uptrend from the macroeconomic point of view often looks at company level like an obstacle race strewn with deathtraps. A company will speed along merrily for a time – long or short – then teeter and drop out, or master the crisis and rise to new heights. The third industrial revolution, repeating many of the experiences of the second in quick motion, tends towards extremes in this respect too.

Very few companies have managed to stay in the premier league of this industry over time. Constant ups and downs are accepted as the day-to-day norm. Nixdorf is a prime example. Long the epitome of a successful European computer company, it only survived into the 1990s due to a takeover by Siemens. In consumer electronics, fine-sounding brands like Wega and Braun have disappeared from the market entirely. Even IBM, the electronics giant that controlled about a quarter of the total computer market at the end of the 1980s and generated almost half the industry's profits, has shed much of its glory, though it still has some excellent divisions.

Others have escaped from under the shadow of the giants. Since the early 1950s, new businesses such as Hewlett-Packard have grown into industry giants themselves, and in turn have had to square up to attacks from even more recent start-ups such as Apple or Compaq.

All turbulence notwithstanding, the industry is clearly divided into winners, who are building their profitability, and losers, whose returns are plunging further and further into the red. The underlying patterns have been confirmed in an international long-term study conducted by McKinsey in association with the Universities of Augsburg and Stanford (the study is summarised at the end of this chapter).

Average returns on sales (ROS) in this sector range from over 20 per cent down to minus 9 per cent, and annual growth rates range from over 26 per cent to minus 8 per cent in individual segments (Figure 1.1). But the four segments (defined in the study as consumer electronics; computers/communications; large systems; and industrial electronics/measurement systems) are not good or bad *per se*. For example, both the consumer electronics and computers segments ran at a loss overall in 1991/92 and so destroyed real corporate value, but both leaders and laggards in profitability and growth are distributed over all four sectors. While the best companies are expanding their lead and widening the gap, the mid-range players are dwindling, and the ranks of the laggards are swelling (Figure 1.2).

More painful, for European companies, is the uneven regional distribution of winners and losers, and in recent years the performance of European suppliers has been particularly poor. Depending on the segment, only one-fifth to one-third of European companies can currently be called successful, and even these will be at risk once the long-standing price shield of regulated supply relationships (common in many sectors) is abolished in many areas. Today, European computer companies meet only one-third of domestic demand, and a large part of 'European' computer manufacturing is in the hands of American and Japanese companies.

The particular dynamism of this industry is causing the losers problems, in Europe as elsewhere. But it is precisely this combination of competitive and innovative pressures that the successful companies

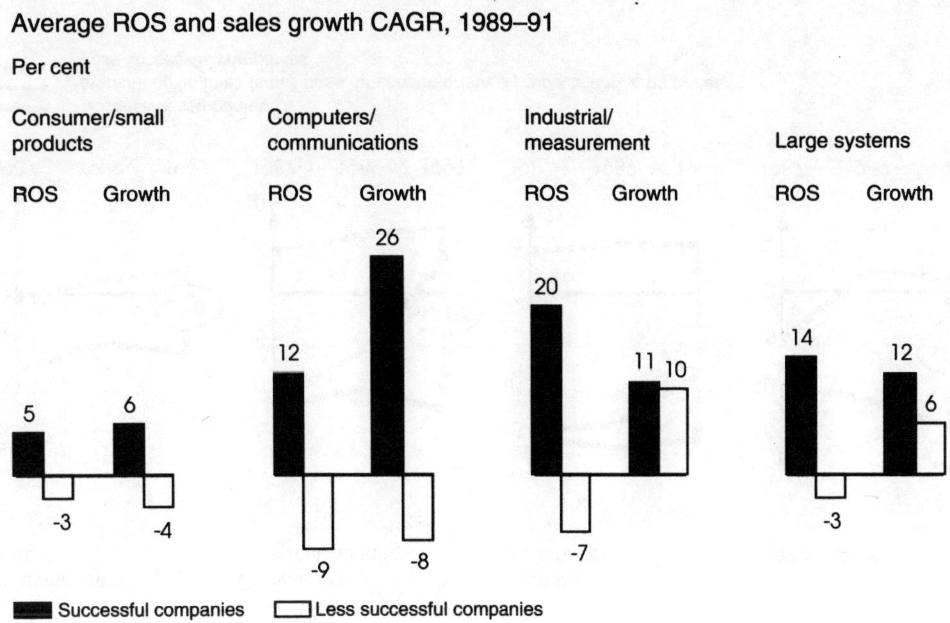

Fig. 1.1 *Surprising performance gaps*

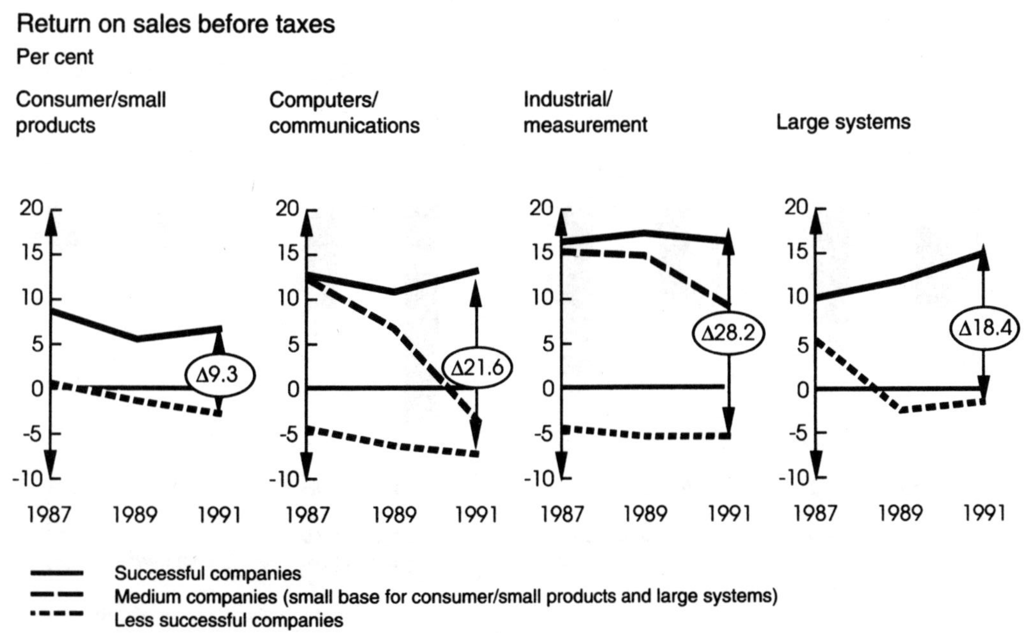

Fig. 1.2 *Only successful companies are maintaining profitability and their lead is growing*

have exploited – not like a machine that is running ever faster and better and more efficiently, but like a living organism that is becoming increasingly intelligent, strong and dominant in its environment.

And in this Darwinian process, failure to adapt quickly means extinction. Getting back on track isn't easy. A straggler starting in 1991 and aiming to reach the top by 1994 would have to increase labour productivity by about 30 per cent each year – an effort equivalent to a full-scale turnaround every year, and a feat which, in our experience, only the most resolute of management teams can hope to achieve. On the other hand, in this environment management positions become easy prey. Even a company with a successful track record cannot rest on its laurels; to stay at the top it has to increase productivity by about 7 per cent a year (Figure 1.3).

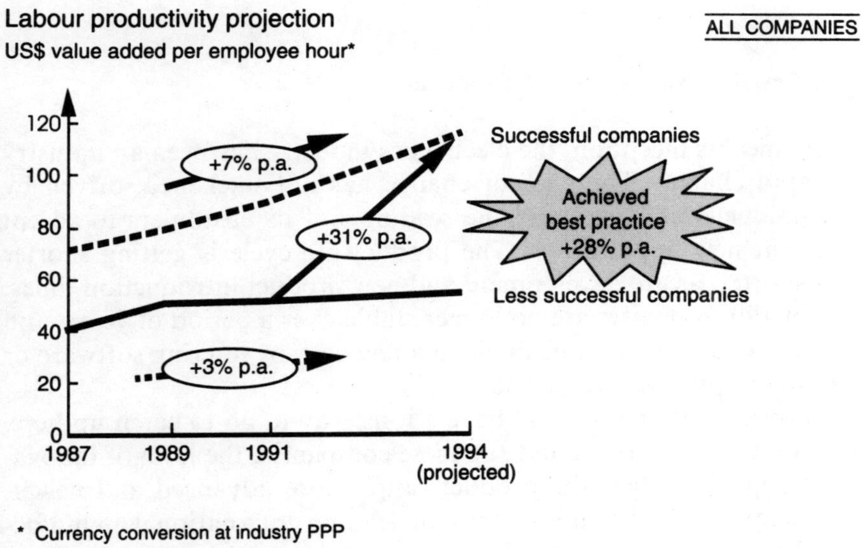

Fig. 1.3 *Less-successful companies need dramatically higher improvement rates to catch up*

To play at the world-class level, it is crucial to understand the game-rules: there are penalty and bonus points in cost, time and quality simultaneously. The world-class league has adopted new rules here, examples of which can be seen daily. The long-term McKinsey study has clearly confirmed this on a global basis.

CLASS DIFFERENCES IN EVERY AREA

The triumphant progress of many products can be attributed to plummeting prices. But the underlying constant productivity increases have a down-side: a loss in value-added that can reach 50 per cent in two to three years. To keep growing, or even to maintain current levels of value added, already requires very aggressive volume growth.

Over the past decade, these improvements took top priority for leading companies. Restructuring, time reduction and quality programmes, together with innovation programmes and aggressive global expansion, boosted performance tremendously – leaving less-successful companies even further behind. Today, these companies are feeling the lack of innovative, globally-marketable products, a competitive cost base, and a global presence.

Too Few Innovative Global Products

Ever since its inception, the electronics industry has been an industry of rapid change. The pace of change has not slackened, driven by technological advances and the readiness of its customers to accept constant new applications. The product life cycle is getting shorter and shorter, as are development and new product introduction times. Production processes are no longer stable over a period of years, and changes can now often be made in a few days by altering software or modifying printed circuit boards.

European companies still have a long way to go to catch up here. In successful American and Japanese companies, the level of innovation is much higher, the product range more advanced and newer. Compared with the innovations of leading international manufacturers of consumer electronics and computers and communications, such as high-definition, low-cost colour printers, flat colour monitors, vest-pocket-size intelligent personal 'assistants', and small portable video recorders, the Europeans have very little to offer.

European computer manufacturers generate only 24 per cent of annual sales from products introduced in the last twelve months, whereas American and Japanese firms get 60 per cent. The gap is the same in consumer electronics: 37 per cent from new products in European companies compared with 80 per cent at Japanese companies.

High Costs and Productivity Disadvantages

Increasing globalization has triggered relentless competition. Companies looking for new sales opportunities are concentrating more on building market share through increased customer value or price advantages. So they pass on to the customer the benefits of productivity increases from innovations and learning-curve effects. As a result, prices are falling faster than ever. The PC that cost $5000 a few years ago is available now for $500, and its performance is better, too. Measuring equipment and computer tomographs cost half the price. Annual price cuts of 10 to 15 per cent are not uncommon in the electronics industry.

Most European manufacturers have not been able to keep up with the productivity advances of the Japanese, though many American companies now have. The labour productivity rate of European companies is currently 25 per cent below best practice in consumer electronics, and at least 40 per cent below in all other sectors (Figure 1.4). If suppliers outside Europe stay as 'aggressive' as they are, the gap will widen. It will be very hard to close it in the short term.

The cost position of many European companies matches their low productivity: cost disadvantages of 30 to 50 per cent are common. Because of high competitive and pricing pressure, inflated costs cannot be passed on to the customer. With dropping prices and slow productivity growth, profits have also steadily fallen. Compared with competitors from other parts of the world, European industry is at the low end of average.

Yet price pressure in Europe is still less than in the highly competitive markets of Asia and the US, where there is a strong correlation between success and labour productivity. Many companies in Europe are still financially successful despite comparably poor productivity (Figure 1.5). But in future, the price premiums of the past will become harder to achieve as the traditional strength of European companies – superior-quality products – ceases to be generally true. On the contrary, defect rates are often much higher than for rival products from Asia and the US.

The situation has to change. Certainly in the emerging environment, where companies from Asia, outside Japan, are catching up in the productivity race, a merciless shakeout of the weaker companies in US industry is taking place, and world market prices are becoming increasingly similar. And what hope is there? Well, although the

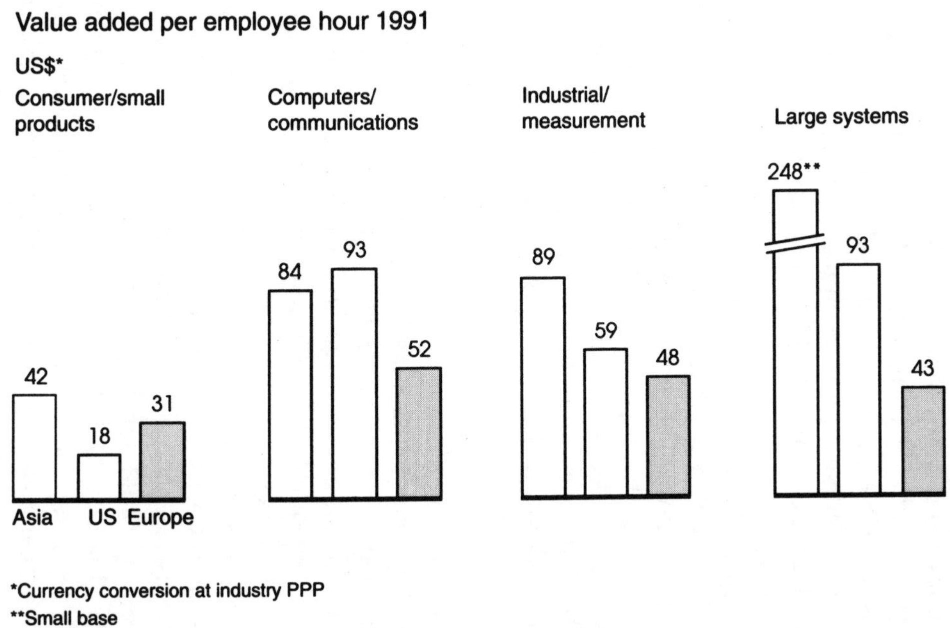

Fig. 1.4 *Much lower labour productivity in Europe*

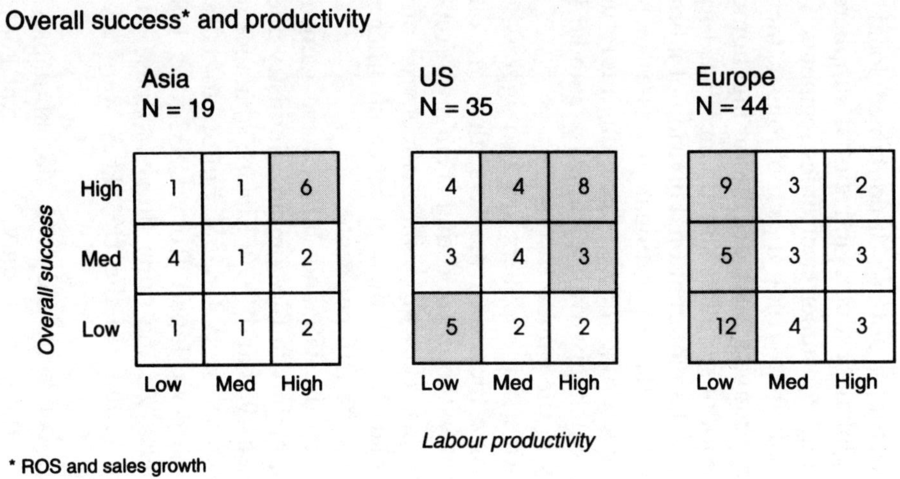

Fig. 1.5 *Europeans' success not necessarily based on higher productivity*

matrix in Figure 1.5 shows most European companies in the 'poor overall success and low labour productivity' box, at least two companies show that this does not have to be the case.

Lack of Global Presence

The electronics sector is not a self-contained local market. It is characterized by global products and global suppliers, particularly in the computers and consumer electronics segments. Internationalization is so far advanced that it is no longer merely customer requirements and products that are becoming homogeneous. An international network of suppliers and alliances has formed, allowing global production and an international division of labour. And it will surprise many to hear that operational processes in product development and manufacturing are also becoming more uniform. Companies unable to catch up with world-class standards will find it impossible to stay in the game.

Japanese and American suppliers are also aggressively exploiting opportunities outside their own markets; they are on the advance internationally, but without neglecting their own domestic markets. Not so the Europeans. They are being manoeuvred out of their own highly attractive market, but their presence in other regions of the Triad is usually less than 10 per cent.

Competitors from the US and Japan have gained a firm hold around the world, mainly in the computer industry and in consumer electronics. Between 1984 and 1990, Europe's balance of trade deficit *vis-à-vis* the US rose from US$5.4 billion to US$10 billion in computers, and *vis-à-vis* Japan from US$3.6 to US$7 billion in consumer electronics. At the same time, Japanese companies were even able to gain ground in the US. They were the real winners during this period.

In the European computer industry, US companies have a dominant 62 per cent of the market; only 28 per cent goes to European-owned businesses. This contrasts sharply with Japan or the US, where over three-quarters of the domestic market is served by local companies. Market shares in individual European countries are even more telling. Only American companies like IBM, Hewlett-Packard, or DEC are equally strong all over Europe. All European suppliers have comparably good market shares in their domestic market, but nowhere else.

Global mass markets have not yet developed in industrial electronics. Individual companies have specialized in profitable applications or regional niches. But with the introduction of open systems, this too will change in the next decade. There is still time to build a strong position.

All in all, a sobering survey for the weaker competitors, and thus for many European companies in particular. The gap must not be allowed to get any wider, or it will be impossible to close. If Europe wants to play a role in the electronics industry of the future, it will have to change direction quickly. Increased innovation and a readiness to change is needed, rather than an insistence on sticking to structures and behaviour that have served well in the past.

NEW GAME-RULES FOR WINNERS

For almost ten years it looked like the ultimate success story: Compaq Computers, founded in 1982, took the American market by storm with its personal computers, setting standards both at home and abroad. The low-cost PCs with MS-DOS operating systems penetrated the office domain of IBM PCs and, thanks to new distribution channels, market shares in key European countries also developed very nicely.

Yet by the late 1980s, improvements in productivity and sales no longer matched up to the fierce pace of the competition. In 1990–91 sales fell by 17 per cent. Compaq, used to being pampered by profits, slipped into the red for a quarter. This was enough of a shock to the company's system – the salutary shock nearly always needed for a successful turnaround.

What Compaq did next in 1991–92 is a textbook example of how to succeed at the top of this dynamic industry.

The founder and chairman left. His successor turned round the previous strategy of moving more and more up market (into workstations and so on) by 180 degrees. The aim was now to attack the mass market with cheaper products: make PCs as cheap as pocket calculators – and as popular.

Nearly every management tool in the book was used. Rigorous design-to-cost made it possible, for example, to reduce the price of a 386 PC by nearly 60 per cent, without significantly reducing customer value. At the same time, over 40 new models were developed and launched in a short time. Production was streamlined, and the number of retailer outlets almost tripled worldwide. With network servers, Compaq exploited a new mass market. Some former managers who did not move with the quick pace of change were replaced.

The high-volume strategy soon demonstrated early wins. In 1992, further operational improvements resulted in a second price reduction of about 40 per cent, followed by a corresponding surge in demand. Two years later Compaq became the market leader in Germany, for example, for laptops. In Japan, the company was even able to launch a successful low price offensive against the domestic PC market leader, NEC.

The net effect was a dramatic increase in labour productivity resulting in 50 per cent growth with 25 per cent less labour. Productivity had thus doubled, but staff reductions were limited to 25 per cent because restructuring had simultaneously triggered new growth.

Less than a year after launching the turnaround, Compaq's drastic remedy put the company back into the profit zone.

This kill-or-cure method can also serve as a model – for leaders and laggards alike, for electronics and many other manufacturing industries. Traditional, incremental improvement programmes have outlived their usefulness – even though many companies, particularly European, may need to catch up in the areas they address. World-class players completed such programmes a long time ago. They know that for true differentiation and resulting improvements in competitive position, these programmes are no longer enough. Even more radical catch-up measures are called for.

Restructuring and growth are the tenor of the new game-rules in the world-class league. This combination of sustained, progressive productivity improvement and innovation stems from a better understanding of customer needs, technological development, and internal processes. According to observations from McKinsey's long-term study and from many years of consulting in the US, Asia and Europe, the game-rules can be summarized as follows. They are described in more detail in the subsequent chapters of this book.

1. The Best Companies Improve More Radically and Faster

Successful companies know that they can only secure long-term growth and employment by continuously raising productivity. Old jobs are shed and new jobs are created for the extended activities. Conversely, the pressure for a radical increase in productivity quite often emanates from tremendous growth – as in some American high-tech companies which have doubled volume within a year. Since it is not feasible to integrate and train enough new employees, opportunities for sales growth drive increases in productivity.

Heel-dragging on change and waiting for better days is the beginning of the end in the electronics industry. Sharp breaks with the past are needed, and quick evolution which, from the point of view of slower players on the outside, bears more resemblance to revolution. The best companies know that they have to sustain the swift pace of change. If they take even a short breather, competitors will pull ahead.

People may mourn the passing of the good old days of stability, but the only route to competitive survival is to play by the new rules and substantially improve mobility and flexibility.

2. Focus for Fitness

Focusing not only creates profit improvement opportunities, it almost invariably generates attractive growth prospects as well. Successful companies, as the findings of the electronics industry survey show, dodge the escape route of apparently attractive and highly profitable, but also highly-fragmented high-end niches. They are aware of the downward spiral of declining volumes and rising costs that entry here will trigger. Instead, they occupy or create new high-volume markets.

With this approach, companies reduce their product complexity, stop producing 'individualized' products, and focus their resources on fewer products and businesses. The result is an increase in innovation and productivity. Unencumbered by the heavy complexity burden, they achieve much greater dynamism and speed. No wonder catching up often has to start with radical complexity cuts.

3. High Costs are Mostly Home-Made

Companies that enjoy high margins over a period of years tend to have inflated cost structures. Compared to the top international competitors, a cost–productivity gap of 50 per cent can easily occur. Closer analysis often reveals that about half of this (25 percentage points) can be attributed to failure to design for manufacturing, overcomplexity and product features not appreciated by the customer. A further 10 to 20 percentage points are due to inefficient processes and organizational structures. In contrast, the share of factor cost differences (for example, different working hours and wage levels) is low at about 5 to 10 percentage points. Internal efforts can thus make up the greater part of the deficit.

How to do this is shown daily by successful companies, even in countries with high wage levels. They have adapted to their specific local conditions with design for manufacture, improvements in operational processes, and adaptations to production 'architecture' – instead of relying on increasingly complex automation. These companies show that the production of high-tech goods does not have to migrate to countries with lower factor costs, since successful restructuring can narrow the difference in personnel costs to little more than the sum of freight charges and customs duties.

4. Successful Innovation is Measured in Yield

Top companies are more than twice as productive in R&D as weaker performers. This is achieved mainly by focused use of resources, the creation of small teams with real 'stars', high flexibility and an excellent workplace infrastructure with excellent equipment.

While less-successful companies are still trying to impose greater discipline on their development processes, the top companies have already gone one stage further. They make many more product modifications just before product introduction thanks to short, fast feedback loops that can push the design freeze point very close to the launch date.

Such companies develop a new PC, for example, in just six months and allow modifications up to about six weeks before market launch. In contrast, less-successful companies need about a year, and have to stop making changes five months before market launch. And successful companies take care not to mistake technological excellence for genuine customer value: they guard against over-engineering by having interdisciplinary development teams. The PC development team, for example, will include people from marketing and finance.

5. New Businesses Must Hit Windows of Opportunity Just Right

High profitability in old businesses leads many companies to neglect new businesses and market opportunities. These companies shrug off the early wins of competitors they once led; they grow complacent and often miss out on opportunities as a result.

Particularly in a dynamic industry like electronics, an innovative product offering is not enough to guarantee a successful new core business. Added to this must be positive feedback from the interaction of the innovations with the environment. This feedback – from an existing (or missing) technical infrastructure for instance – accounts for the fact that new products can usually only be successfully marketed within a very narrow time frame. It also explains why – from virtually identical starting points – one innovation will be a hit, and a similar one a flop.

6. Nothing Fazes the Problem-Solving Organization

In successful companies, functional organization is a thing of the past. The change has been triggered by mounting complexity that could not be handled across interfaces, in conjunction with growing demands for rapid-response capabilities – colourfully expressed by one foreman as, 'We're not shooting at sitting ducks, we've got charging bulls here!'

Other factors that have helped break down functional structures are better training and a highly skilled workforce, fully capable of tackling complex problems holistically. Successful companies know that they have to set macro goals, such as profit targets, for their new products. At the same time, they avoid detailed instructions and micro-management of individual parameters that quickly become out-of-date and which, under discussion, easily become the subject of whitewash jobs.

Successful companies manage to infuse the entire organization with a philosophy of continuous improvement, and they teach each employee to live this philosophy every day. It is impossible to faze these problem-solving organizations even with high rates of change, because they are small and flexible rather than large and unwieldy, integrated rather than fragmented. Entrepreneurship and a willingness to take calculated risks take the place of command and control structures.

Less-successful companies, however, believe their organizations would not be able to cope with radical and rapid change, and fall further behind.

7 A Catch-Up Strategy Must Combine Restructuring *and* Growth

We know how successful companies reach the top and stay there. To catch up with them or overtake them requires a willingness to make major, sustained efforts.

Cost-reduction programmes are nearly always the first step towards closing the gap. Savings can be precisely quantified and realized, successes can quickly be seen. Successful companies know, however, that they can only maintain positive motivation in the long term if they give their employees back the experience of winning. Reason enough to develop growth programmes in parallel with restructuring programmes.

The 'shock of the gap' provides the stimulus to carry out determined restructuring by creating small, empowered organizational units, lean product lines, and low-cost products that give end-users no more and no less than the value they are ready to pay for. Streamlined core processes, such as new product introduction from development to market entry, ensure lasting gains in efficiency. All this creates the momentum for fresh expansion: with existing products and customers, in new regional markets and customer groups, with the new products from a rejuvenated product range.

It is possible to make a comeback from a near hopeless position even in the electronics industry, as demonstrated by examples like the recently publicized case of Siemens telephones. Within three years, a dying division was transmuted into an internationally respected competitor, now able to compete successfully against competitors from the Far East with simple 'commodity' products made in Germany.

These and other successes are a never-ending story. The only way to avoid the downward spiral is by constantly developing and launching new products with high value-added, while constantly making further drastic improvements in productivity. There are no timeouts in this game.

* * *

The extreme dynamism of the electronics sector with its high competitive pressure and breathtakingly swift innovation cycles cannot frighten successful companies; their adaptable and fast-learning organizations give them a competitive edge that is hard to imitate. And the prospects for building this advantage further are good. These companies have adjusted to the pace of the industry and are, indeed, driving it further.

Top performance is possible in every region of the world. However, a top position cannot be reached by hiding behind protective walls or subsidies, or by making tentative efforts at improvements in cost, quality or speed. The only way up is to follow a completely new set of game-rules that add up to aggressive growth founded on sustained, superior productivity improvements and innovations.

THE McKINSEY LONG-TERM STUDY ON THE STATE OF THE INTERNATIONAL ELECTRONICS INDUSTRY

Conducted by McKinsey & Company in collaboration with the University of Augsburg, Stanford University and Stanford Japan Center, the study looked at the electronics industry worldwide and aimed to establish the success profiles and reasons for performance differences between companies, segments and regions. This involved analyzing developments between 1987 and 1991 and forecasts up to 1994.

The survey on which the results are based is currently being repeated to keep track of the rapid changes in this dynamic industry.

Scope of the Study and Regional Coverage

The survey polled 102 companies in the countries of the Triad – 45 in Europe, 34 in the US, and 23 in Asia. The companies were drawn from the four sectors of: industrial electronics/measuring systems; large systems; computers/communication; and consumer electronics/small products (see Figure 1.6). The study did not look at major companies in their entirety, but surveyed business units or individual companies and the main product group within each of these business units or individual companies. Component manufacturers, producers of military electronics and software houses were not included in the survey.

Sector Coverage

The companies surveyed cover between five and 28 per cent of world sales in their individual sub-sectors. The computer sector respondents include five of the top nine mainframe computer manufacturers, five of the top 10 for medium-sized systems, six of the top 15 for PCs and three of the top five for printers. Coverage of the other three segments is similarly wide. For example, the database includes figures for four out of the top six television manufacturers and two of the top four manufacturers of control and regulatory systems. This wide coverage enables us to draw reliable conclusions about the situation in the industry.

Approach

Information on the success patterns of the leading companies and the behaviour patterns of less-successful companies was drawn from the evaluation of a detailed questionnaire and from detailed on-site interviews and a wide-ranging discussion in participant conferences, plus the experience gained from McKinsey consultancy projects. To formulate valid conclusions on an international basis, non-currency-based parameters (for example, R&D expenditure as a percentage of sales) were used in most cases. Where monetary values were unavoidable for parameters such as value-added per employee, product-specific purchasing power parities based on market prices were used to convert the various currencies. This procedure gives better results than a conversion based on exchange rates.

Corporate success was defined using a 'success index' which made it easy to see how individual companies compare with one another in absolute and relative terms. The index gives equal weighting to the average return on sales and the average growth in sales for the years 1989 to 1991.

In ranking the companies, the different levels of return and growth in the sectors studied were taken into account. For example, a top performer in the consumer electronics/small products sector may be on the same level as a top performer in the computers/communication sector, although the latter has done much better on absolute return and growth figures. Plausibility checks using other, more complex indices (such as cash-flow parameters or ROI) confirm the rankings given by the success quotient, except for marginal position changes, leaving the conclusions unchanged.

The main driver of corporate success proved to be labour productivity. Our productivity definition refers, unless otherwise stated, to the unit 'gross value added' (sales less materials less stock changes) per hour of work. Varying levels of vertical integration and different annual working hours have thus been adjusted for; individual price premiums for particular companies must still be allowed for when making specific comparisons across various markets and countries.

The conclusions of the long-term study have been verified in many discussions with participating industry managers, academics, and consultants. Numerous individual examples in the public domain further reinforce the correlation observed between patterns of conduct and business success or failure.

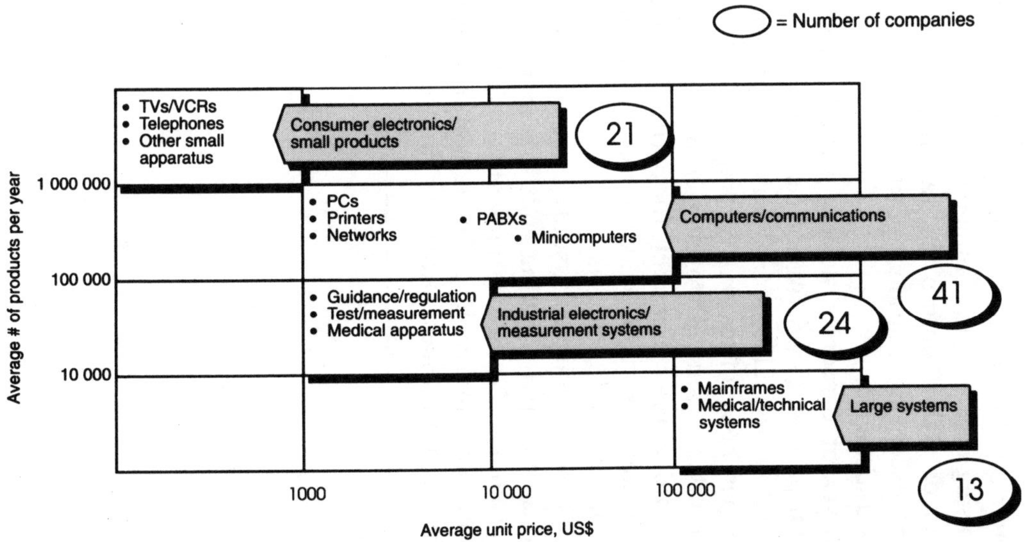

* Of which 3 had no specific segment orientation

Fig. 1.6 *The survey covered 102* companies from four industry segments*

2 The Best Companies Improve More Radically and Faster

In an industry where fortune and famine are such close neighbours, the stars of today may well be the dropouts of tomorrow. All it takes is one mistimed entry into a new market, or a productivity increase of just 5 instead of 10 or even 15 per cent for a very few years in a row. Aware of the fatal trap of doing 'too little too late', the best companies put every effort into improving continuously, faster and more radically, in both restructuring and growth.

The best companies have mastered the 'V' concept. The association with 'victory' will occur in this context, which is not far wrong. But the 'V' has found its way into McKinsey vocabulary largely because of its shape, symbolizing the dip of restructuring and the ascent of growth. The depth of the 'V' is not an absolute parameter; it varies by segments and industries. The 'V' of the electronics industry is steeper than that of the metalworking industry, for example, because change processes such as product line revamps happen faster in electronics. Even between segments in the electronics industry, there are huge differences in the rates of change in restructuring and growth, and thus in the depth of the 'V'.

But one universal factor is that successful companies have a more radical approach to the 'V' concept than their less-successful competitors. Their restructuring is more sweeping, their growth is more aggressive, and their approach to exploiting the interaction of the two is virtuoso.

SWEEPING RESTRUCTURING, AGGRESSIVE GROWTH

Our long-term study showed that many companies underestimate both the need for improvement and the speed at which they must improve. Some swear by restructuring *or* growth. Others try both, but with differing degrees of conviction. If you look at return on sales, however, you will see that the most successful companies follow both tracks with particular vigour (see Figure 2.1).

- The *pure expanders* are on the growth track only; they give it higher priority than restructuring. Despite sales increases of 75 per cent, their return on sales fell slightly between 1989 and 1991. They do not exploit scale economies systematically enough, do not get slim enough, and thus cannot react fast enough to changing conditions.
- The *pure restructurers* are at the other extreme. Over the same period of time, they managed to increase their return on sales from 1.4 to 6.2 per cent. Quite a success, but only sustainable if backed up by an active expansion programme. If sales are falling, on the other hand, not even the biggest restructuring programme will increase productivity as much as is needed to guarantee profitability in the long term. This is the painful experience of any company that has implemented a sweeping restructuring programme but was robbed of the hoped-for revenue turnaround because of collapsing sales. Trying to jack up productivity when business is declining is a hard and doubly-exhausting battle that can easily end in mass layoffs, and often in revenue and cash-flow problems.
- This is a problem for the *gradual improvers* – a third of all companies analyzed – who are losing out all along the line and have not understood how to capitalize on a good starting position. In two years they improved their cost position by only two points, and their productivity actually declined. This diminished their prospects for growth. They have made little impact outside their core business, and even there they have lost entire segments. The result was a slump in ROS from 10.2 to 4 per cent – from the top to the bottom of the ratings in just two years.
- The *improvement champions* are quite different. With annual increases of 8 to 14 per cent in productivity and 21 to 48 per cent in sales, they attained above-average success in both restructuring and growth. The resulting return on sales of 7.7 per cent is an excellent figure, well above the starting point of 4.3 per cent.

The critical factor is the rate of change or the annual depth of the 'V' (Figure 2.2). But the achievement target a company has to set

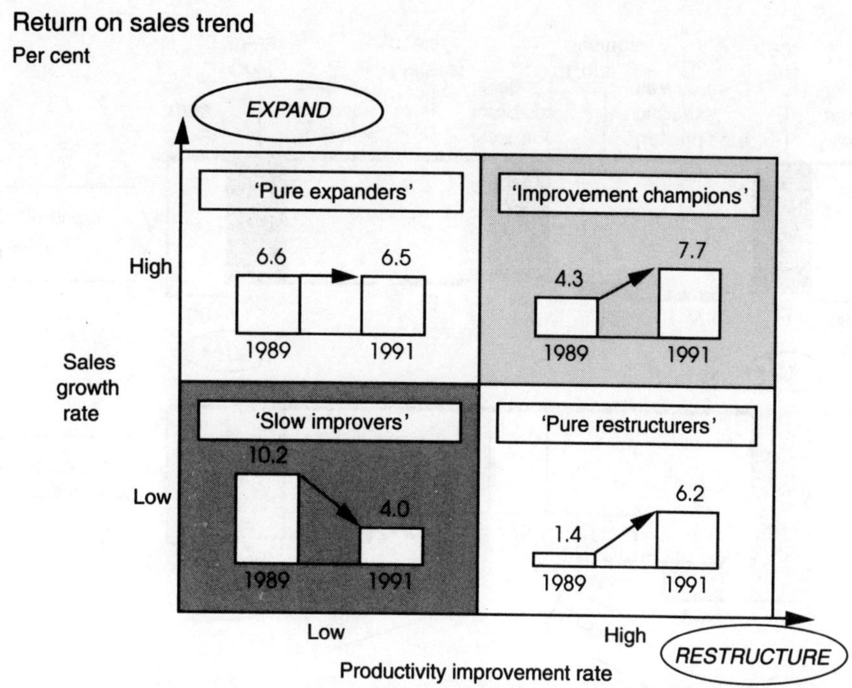

Fig. 2.1 *Superior profitability through aggressive restructuring and growth*

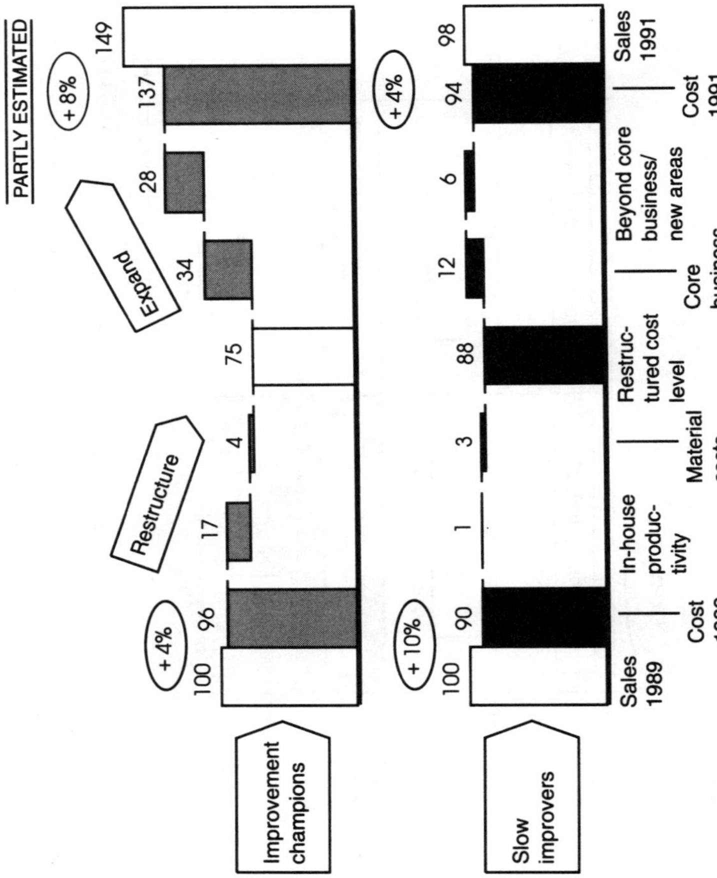

Fig. 2.2 *Improvement champions have deep 'V'*

itself will largely depend on its current level of performance measured against the best, and on the expected future annual performance improvement of the best.

European companies often found that their productivity was still quite competitive in the early 1980s. But in the years that followed, they raised it by only 3 to 6 per cent per annum, compared with 8 to 15 per cent by their competitors in Japan and the US, leaving them faced today with a productivity gap of easily 50 per cent. Added to this, they lost market share worldwide to faster-growing US and Asian competitors.

Ideally, improvements on one side should stimulate progress on the other. The most successful companies are those with the best growth rates.

GROWTH THROUGH RESTRUCTURING – AND VICE VERSA

The pace of improvement in the electronics industry is set by the rapid advances in performance that lead to tumbling prices and new or enhanced applications. The close interrelationship between restructuring and growth is repeatedly demonstrated in this process (as shown in Figure 2.3).

For instance, creating new businesses with new products or expanding into new regions requires considerable resources, not only in development, manufacturing and sales but above all in top management. These resources can often only be released by discontinuing yesterday's businesses and product lines in good time.

The best companies pursue this kind of product portfolio design strategy quite systematically. Their research and development staff do not go on optimizing the old familiar products to the crack of doom, and their accounting system does not block the transition to the future by pointing to the supposed profit contributions of the old businesses, but is ready to finance the risks and opportunities associated with new products and ventures. Their top management knows what would happen otherwise: their competitors would get to the market first, cream off the profits, and a lot of cash and energy would be needed to start a new catch-up effort.

The relationship between shrinking and expanding is similar in product design. Design-to-cost programmes are often the most powerful lever in reducing product manufacturing costs. Reductions

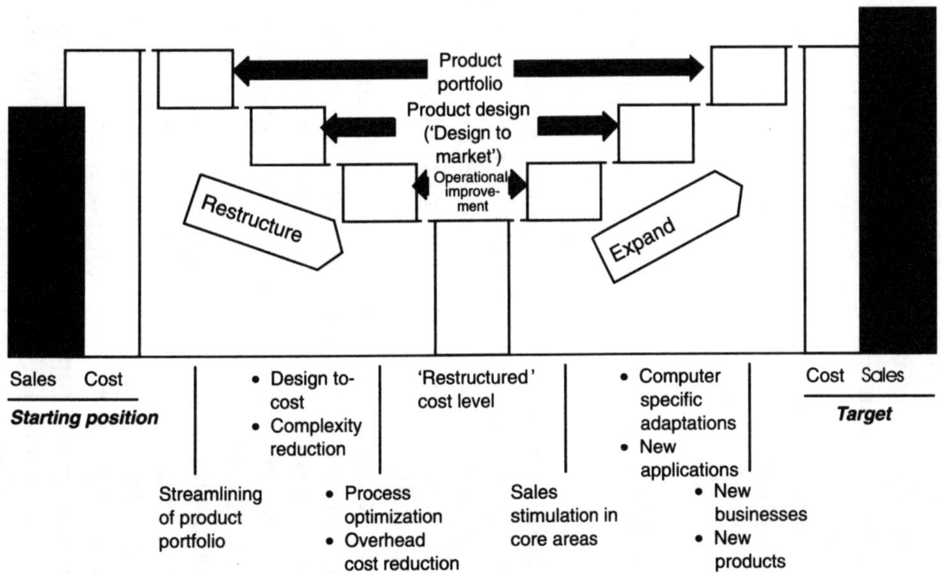

Fig. 2.3 *Restructuring and growth are closely linked*

of 50 per cent every three years are not unusual. Whether it is wise to pursue such goals or not is no longer at issue. The question, rather, is who will be first to do it and pass on the benefits to customers in the form of lower prices. The pioneer will capture the growth that produces economies of scale in fixed costs, leading in turn to further productivity increases and so to price reductions in the market – an upward spiral for the winners, a downward spiral for the losers.

The rapid fall in unit cost can be converted into spectacular rates of growth, particularly in businesses with high price elasticity. Cost-optimized products also make it possible to open up new applications and adapt products to individual countries. But, again, this success spiral only works for companies that take an aggressive approach to restructuring and growth.

Practically any electronics product can be used to demonstrate how restructuring and growth increase productivity and become visible in the market in the shape of falling prices. Mobile phones, for example, cost over DM 4000 in Germany at the start of the 1990s. By 1994 they were available for under DM 300, and if the buyer subscribed to a network operator at the same time, for less than DM 50. As prices fell, sales soared. This may seem a particularly extreme example, but annual price declines of 10 to 15 per cent are quite common in the electronics industry, and the trend is unlikely to change.

Technological advances also destroy an enormous amount of value-added: functional groups contain fewer, but more highly integrated components; software replaces hardware; vertical integration is reduced. Production time therefore shrinks drastically despite an increase in unit volume, and entire production lines and even locations are no longer utilized. It is not always possible, as in the case of the mobile phone in Germany after the 'D' and 'E' networks were introduced, to compensate for the loss in value-added with increased production volume. Enhanced applications such as telephones with fax, or PCs with networking capabilities are opportune new arrivals in this situation, offering the chance to boost volume and value-added.

And this closes the circle. With vast growth opportunities, the new applications often require new sales channels, new skills on the part of developers, and new corporate processes. Companies must constantly monitor the use of their financial, staff and management capacities, reallocate resources, channel manpower, money and managers towards new growth opportunities and, therefore, pull them out of old businesses.

Growth through restructuring, restructuring through growth – the relationship is self-evident. The more extensive the restructuring, the greater the growth prospects; the stronger the growth, the greater the need or indeed the potential for restructuring. This basic pattern applies equally to entire conglomerates operating in many different businesses and to individual business units, to major companies and to mid-sized companies, as the following examples show.

General Electric

An impressive corporate restructuring and growth programme is described in the book *Control Your Destiny or Someone Else Will*, a report on General Electric's rise to the premier league of leaders in growth and profitability.*

The turnaround began in 1981, when Jack Welch became chief executive. In ten years sales rose from US$27 to US$60 billion and ROS from 6 to 7.5 per cent. GE's *Forbes* ranking soared from no. 11 in 1981 to no. 3 in 1992. Its stock price increased five-fold over the decade.

What is the hidden secret behind the outwardly visible signs of improved economic performance? An immense effort in reallocating internal resources, a classic example of simultaneous radical slimming and growth.

In the space of ten years, Jack Welch sold off parts of the company with sales of US$11 billion. The only businesses he kept in his portfolio were those where General Electric could become no. 1 or no. 2 in the world market. At the same time he initiated efficiency enhancement programmes which cut no fewer than 100 000 out of 400 000 jobs between 1981 and 1985. All the cost levers were pulled at the same time: overhead costs, throughput times, inventory, logistics, quality, automation and so on. He also flattened the management hierarchy from an original nine levels to between four and six.

This restructuring programme was accompanied by a variety of sales-boosting measures. The acquisition of new businesses alone brought turnover of US$26 billion.

---

*Noel M. Tichy and Stratford Sherman, HarperCollins, London, 1995.

Will General Electric be able to sustain its swift rate of improvement in the long term? Jack Welch himself says: 'My biggest mistake by far was not moving faster'.

General Electric will undoubtedly stick with its combined programme of restructuring and expansion. The company, like many others, has learned from experience how ineffectual one-sided attempts at optimization can be. When it concentrated exclusively on restructuring between 1981 and 1984, annual productivity improvements (inflation-adjusted) increased from 1 per cent to 2.5 per cent, but sales stagnated at about US$27 to US$28 billion. It was the expansion in sales from US$28 to US$48 billion between 1984 and 1987 that marked the breakthrough: in the form of annual productivity gains (inflation-adjusted) of 5 per cent, and even up to 8 per cent in such areas as medical engineering.

Hewlett-Packard

General Electric's success in reshaping its entire organization is paralleled by Hewlett-Packard's redesign of its PC division. The cuts were just as radical. And the result was equally dramatic.

In 1991, Hewlett-Packard sold 70 000 PCs, was no. 17 in the industry and was running at a loss. Two years later the sales figure had risen to over 600 000 units. At the beginning of 1994, sales in HP's computer division were higher than those of Digital Equipment, making it the second biggest American manufacturer of PCs.

This impressive success was the result of holistic change in all dimensions of the business, driven by a new management team that set exceptionally high targets for development time, price levels, profitability and market share and made these targets stick. The emphasis is on the word 'holistic': Hewlett-Packard downsized its workforce by 40 per cent, pared the number of manufacturing locations from 12 to two, and eliminated excessive price premiums of 30 per cent. At the same time, slow-moving products were taken off the market and development times shortened from 18 months to six. Fewer parts were needed for the new, manufacturable designs, where the number of screws in one product, for example, was reduced from 50 to zero. The manufacturing time for a PC fell from 25 minutes to four.

The business was reorganized into three business units to cover different market segments. And an effort was made – successfully – to generate real innovations. The sub-notebook computer and infra-red local area networks are examples of such new businesses.

Chairman and CEO Lewis Platt neatly describes his recipe for success as 'pre-emptive self-destruction and renewal' – simultaneously forced restructuring and growth.

WISI

The dynamism for which the major global companies have to gear up is just as essential to mid-sized companies such as WISI in south-west Germany, a family-owned maker of antennas since 1926.

When the extremely successful core business of terrestrial antennas was substituted by satellite antennas at the end of the 1980s, the dynamics of the business changed fundamentally. New sales channels were created with the new hypermarkets for electrical goods, while dozens of new suppliers attacked the market with satellite decoders and receivers from the Far East. Product life cycles shrunk from five to ten years to about one year, and the value-added process shifted from mechanical parts to electronic subassemblies. In short, technological advance transformed a relatively sluggish business into an exceptionally nimble one almost overnight, with prices falling by up to 20 per cent a year.

And even rising sales (particularly driven by German reunification) could not halt the profit slump. In 1994, owner Frank Sihn and his son Axel embarked on a sweeping programme of corporate renewal, aimed at halving product costs. The product portfolio was completely overhauled, 80 per cent of all products were dropped and products with high future potential were added. The remaining products underwent sweeping redesign: manufacturing costs were to fall by 30 per cent despite an increase in customer value.

Building on a substantial reduction in product and parts variety, management began a complete overhaul of location structures, processes, functions, and also of organizational structure to make the company lean and fast. One of the four management levels was scrapped and the number of managers was reduced by half. Materials management was decentralized, fragmented warehouses were brought together and quality control was directly assigned to the production departments. Purchasing reduced the ranks of its suppliers by more than half, and material specifications were revised and consolidated.

The integrated restructuring programme pushed down unit costs by over 20 per cent within the first year. The remaining workforce was far from demoralized by these tough measures, however, because they realized that the company was not only turning around in double-quick time and sharply increasing revenues, but was even capable of anticipating product and price developments instead of merely reacting to market dynamics.

A first major step has been taken. Whether the present successes in restructuring and growth that have put the company on the 'electronics track' will be sustainable depends on whether the renewal programme is a one-off event or can be established as an ongoing process. The example is encouraging because it shows that even mid-sized electronics firms can succeed from a base in Germany.

* * *

But these cases leave no room for doubt that success can be achieved only through constant reduction and destruction of the status quo in favour of renewal and the future. Progress is not made here in small steps, but results, in Schumpeter's words, from the 'creative destruction' of the cosy old order. Permanent revolutions, a refusal to slacken off after success, and tireless active reshaping of the market – these are success patterns that make sense, but in practice are not at all easy to put in place.

To avoid losing momentum and to ensure that the constant redirection of resources is successful, successful companies have developed a special corporate philosophy. They make no compromises. They create a pro-revolutionary environment. And they constantly reinforce their demand for permanent revolution – not least with their motto 'improve or move,' which applies mainly but not solely to management.

SUMMARY

Success in the electronics industry is to some extent self-consuming and has to be re-earned over and over again. Productivity gains achieved by extreme exploitation of new technologies and learning-curve effects are largely passed on to the market in the form of lower prices. The resulting loss in value-added must be made up for by the individual company increasing the *unit volume* and constantly generating new products and businesses.

Future success therefore depends on sustained rapid change. Fast productivity gains *and* fast expansion are prerequisites for success.

- For companies aiming to close the gap to the best, notably most European competitors, this means an integrated programme of rapid improvements in their business systems, product designs, market coverage and business portfolios.
- Successful companies are not distinguished from less-successful companies by the levers they – constantly – apply, but by their radicalism in doing so. Not content with evolutionary adaptations, they undergo revolutionary transformation over and over again, and have recognized that 'You can't cross an abyss with small steps'.

3 Focus for Fitness

Enough is as good as a feast. But where does the optimum lie? In the electronics industry, where both sprint and marathon skills are vital, ballast takes many forms: product features or whole products, customer groups and regional markets, steps in the value chain or interfaces within internal processes. The best owe their superiority largely to simplicity – for them, restructuring as the elixir of growth also contains the ingredient of intelligent focusing to check this overflowing complexity.

The trend towards greater complexity marked many decisions in the European industry of the 1980s: the widest possible choice of products to meet all customer needs, lot sizes of one in manufacturing, the development of extremely flexible, highly interlocked, extremely complex and expensive manufacturing systems. It did more harm than good, for there was a price to pay for high complexity.

Manufacturing process stability suffered from excessive complexity and a wide choice of variants. As a result of increasing product complexity, development costs also rose, not with linear but with exponential growth. Doubling the number of lines of code in a software product (all other things being equal) results in development costs that are not twice but eight times as high.* Built-in complexity also led to higher costs in later stages of the value chain, in product improvement and management and in sales and service.

But product and assortment variety are just two among many types of overcomplexity, with higher costs just one of the negative consequences. In fact, the complexity trap is itself much more complex than that. Excessive variety, whether in the product portfolio, the internal value chain, or in marketing, is not only expensive, it also offers a potential fragmentation of efforts that can easily distract attention from important issues and lead to significant performance deficits.

---

* See Putnam, L. H., 'A General Empirical Solution to the Macro Software Sizing and Estimating Problem', *IEEE Transactions on Software Engineering*, July 1978.

The danger of 'too much and not enough' is ever-present. In response, the most successful electronics companies make restructuring and growth a permanent part of their agenda. The rule is 'focus in every area – and when you've finished, start again.'

THE COMPLEXITY TRAP: TOO MUCH AND NOT ENOUGH

Superiority through focus at successful companies is particularly evident in their product portfolios. They manage with far fewer products than their weaker competitors. In consumer electronics, such companies need 20 per cent fewer variants to sell $100 million's worth of products. The gap is twice as large in computers/communications, and even larger in industrial electronics/measurement (70 per cent) and large systems (95 per cent) (see Figure 3.1).

A look at the impact of an excess of variants and products across the whole business system shows what top companies gain through focus:

- *Research and Development:* The fewer the projects that development engineers work on at any one time, the more productive they are. If they cannot focus because of too many projects, everything slows down. Having to constantly 'get under the skin' of a different project leads to perceptible and measurable efficiency losses. And developers may also lose sight of the big picture; optimizing a wide range of old, familiar projects may lead to neglect of important and, above all, new areas.
- *Production:* In production, these large numbers of components not only have to be assembled, but stored and transported as well. Added to this are longer set-up times due to more frequent retooling, more quality checks, and higher reject rates (for example, because parts/components get confused). It is almost inevitable that equally complex production control processes, interlocking manufacturing plant and warehouse equipment will be installed and operated at considerable expense.
- *Marketing and Sales:* These functions are generally the drivers of greater product complexity, but can very easily also fall victim to it. Sales staff who have to search through over a thousand order numbers in some cases for one software package and find explanations for long, boring lists of price increases, are not left

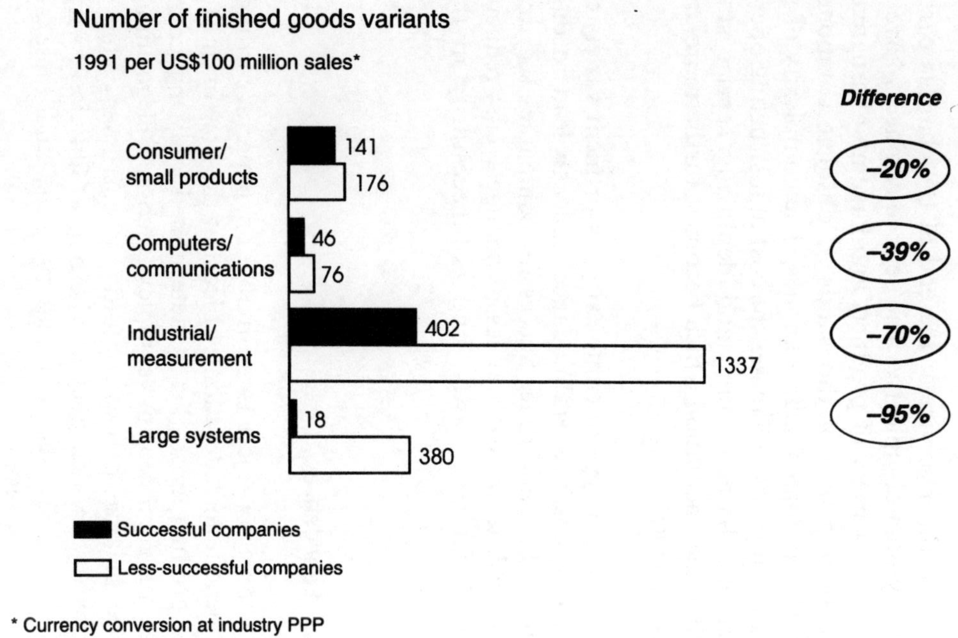

Fig. 3.1 *Successful companies have far fewer product variants*

with enough time to spend on attractive products and markets. Selling time is also sacrificed to acquire the necessary knowledge about a large number of complex products. Extensive product ranges distract attention from the individual product. From a certain number of variants onwards, 'economies of scale' quickly become 'diseconomies'.

- *Service:* Customer service at the end of the chain is particularly hard hit by high complexity. The service engineers have to look after the wide variety of products and have no alternative but to keep on storing tens of thousands of obsolete components for hundreds of outdated subassemblies. The complexity of product design and range also increases the cost of instructing operators at the customer base, training and deploying service staff, and producing documentation which becomes, furthermore, unwieldy and error-prone.

And like every other type of complexity, assortment variety conceals a double risk: a costly overabundance that can lead to dangerous deprivation at the same time. Successful companies see this as an unacceptable risk to their dynamics of change – cost reduction and innovation go far too slowly, and not necessarily in the right direction.

Costly Over-Abundance

Every development project is profitable on paper according to the usual cost–benefit considerations, at least at the time of drafting. If a company's product portfolio is determined only by considering incremental marginal returns, complexity costs will go unnoticed – the requisite balance between additional customer value and an added complexity burden goes by the board. On closer inspection, the big hits subsidize the marginal projects, and these are only attractive on the surface.

European industry has often pursued this fatal course in recent years. Love of technology and misunderstanding of customer needs have lured producers into the complexity trap again and again. Expensive product features have been provided which did not give the customer the corresponding extra value, never reached the awareness threshold and were therefore rewarded poorly or not at all. Driven by technology, new products were developed at great expense for the smallest of niche markets, even though it was clearly unlikely that the potential volume would be big enough to cover costs.

The customers' decision was clear: they turned their backs on expensive, overbred products and moved to cheaper alternatives. These, tailored to the high-volume segment, proved adequately 'spec'd' for most purposes.

This lesson was learned recently by the German machine tool manufacturers, and with them the suppliers of their numerical controls. Having caught up with their Japanese competitors on technology, they believed for years that they would be able to gain the ascendancy with technological superiority. They kept upgrading their machine-tool designs, and increased the accuracy, speed, and number of planes in which the machines could operate. This process was accompanied by cost increases which they had to pass on to the customer. They realized much too late the risk they were running.

The Japanese had not only caught up with the Germans, they had overtaken them. Their products met the technical specifications of the high-volume market in accuracy, life span and flexibility to the entire satisfaction of the customers, and at half the price. Industrial investors welcomed these products with open arms during times of recession when funds were low. They switched to the reliable products of the Far East. The volume business was firmly in the hands of the Japanese, and the Germans were left mainly with the high-price niches for special applications.

Although superficially an attractive idea, securing capacity utilization by covering a wide range of niches rarely works out in practice. The machinery makers and their electronics suppliers in the above example sacrificed major sales opportunities by following this policy, and landed themselves with huge complexity and complexity costs. And they are not alone – which will come as no surprise to those familiar with companies with an engineering culture. It is no coincidence that the average number of parts required by European manufacturers for sales of $100 million is well above the level of US or Asian competitors in all segments of the electronics industry (as shown in Figure 3.2).

Reverse engineering carried out on European products shows that almost every product incorporates many features that the customer will not pay a cent for and that push costs high. For example, one control unit was found to have a cost-reduction potential of 45 per

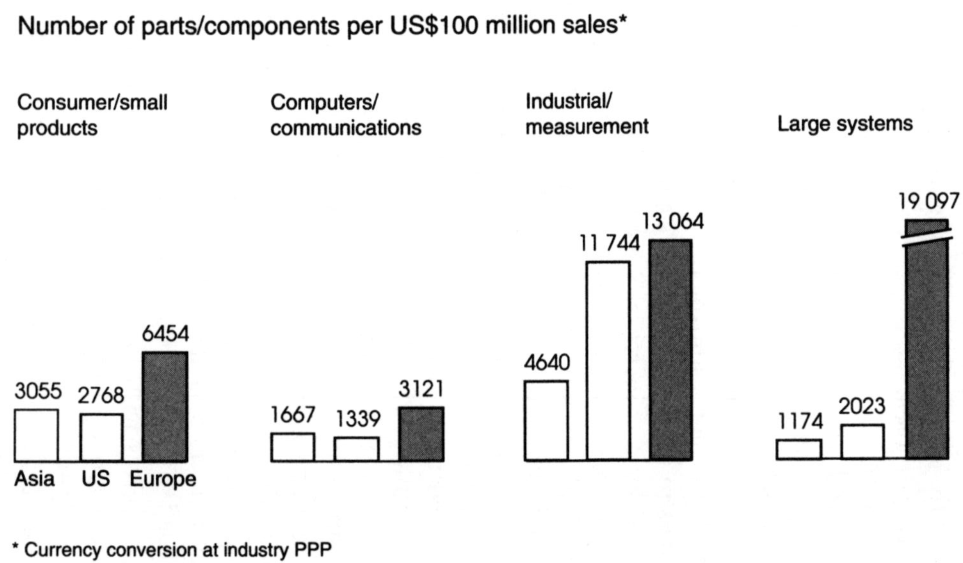

Fig. 3.2 *Large numbers of parts in Europe*

cent (from redesign alone, without adding the potential of a much needed productivity increase in manufacturing). Almost half the savings were on product features with no perceptible customer value. Just leaving them out lowered production costs by a fifth.

The experience of two manufacturers of analysis machines for process control in the chemical industry shows the profit impact of different ways of handling such potentials. With these machines, the measuring procedure varies according to the application. An explosion-proof version is used in most cases, but a very few applications do not need this feature. Single analyses, double analyses, and analyses with a temperature programme are needed to determine specific chemical components.

The largest market segment, with over 60 per cent of process applications, is single analysis in explosion-proof environments. One company concentrated on this volume segment with great financial success. Another manufacturer, not content with one product series, developed no fewer than five, none of them designed for the high-volume segment. But they were all equipped with expensive product features which could only be used in small market niches. High development costs, high manufacturing complexity, and the low volumes achievable with these expensive machines plunged the company deep into the red.

Diverted attention caused by lack of focus, and the associated ballast, is a luxury no-one can afford in the electronics industry's climate of rapid change – and cost is not the only reason.

Deprivation in a World of Plenty

Unit cost disadvantages caused by low-volume variants and over-engineering – a surplus of product features without the value differentiation needed to earn premium prices – are bad enough. But lack of focus becomes fatal when important market opportunities are not covered despite high variety; and this is often the case.

This was the experience of one manufacturer of electronic controls for industrial applications. After the company had made heavy losses because it was unpopular in the market and sales volume had consequently slumped, it reviewed its portfolio of five product series. It analyzed the relevant customer value dimensions, such as the accuracy needed and the number of parameters to be controlled, and finally defined three core segments. However, all attempts to assign the five product series to these three segments failed: together, the series had all the required product features and more, but was packaged in a way that did not exactly match the individual customer segments. None of the segments could really be well served.

In addition, each one of the product series had very expensive features which were only important in tiny market niches and which made the products expensive. In short, the five product series offered too many product features, but not the right ones for the individual market segments.

Redesign was unavoidable. To disbar expensive product features and ensure competitive prices in all product series, the manufacturer decided to stop serving the entire market down to the tiniest niche. Instead, it concentrated on the core customer segments and designed three new product series to replace the five outdated ones. At the same time, the company introduced a modular system based on the low-cost computer hardware supplied by a large-scale manufacturer. This allowed the control units to be fitted with the appropriate computer modules as required.

The new, clean-cut product series reached more than 80 per cent of the market, and the 'expensive' niches were left to others. Better market coverage was attained with fewer than half the number of products and at much lower cost. As a result, sales increased by over 40 per cent within two years and market share grew at the same time.

Experience shows that substantial costs can be avoided by forward thinking, at no quality and market risk, once a company has come to understand the relationship between design complexity, product range complexity, and cost. It is important to find the optimum fit of product series to customer needs. Market segmentation by customer value dimensions and not by technological dimensions or even available product series always makes sense. The less attention management pays to complexity, the more it will spread in product design, product range, production, spare parts and sales.

And this leads us to the escape route from the complexity trap.

THE WAY OUT: FOCUS IN EVERY AREA – AND WHEN YOU'VE FINISHED, START AGAIN

Successful electronics companies owe their ability to win and hold the leadership in both restructuring and growth largely to uncompromis-

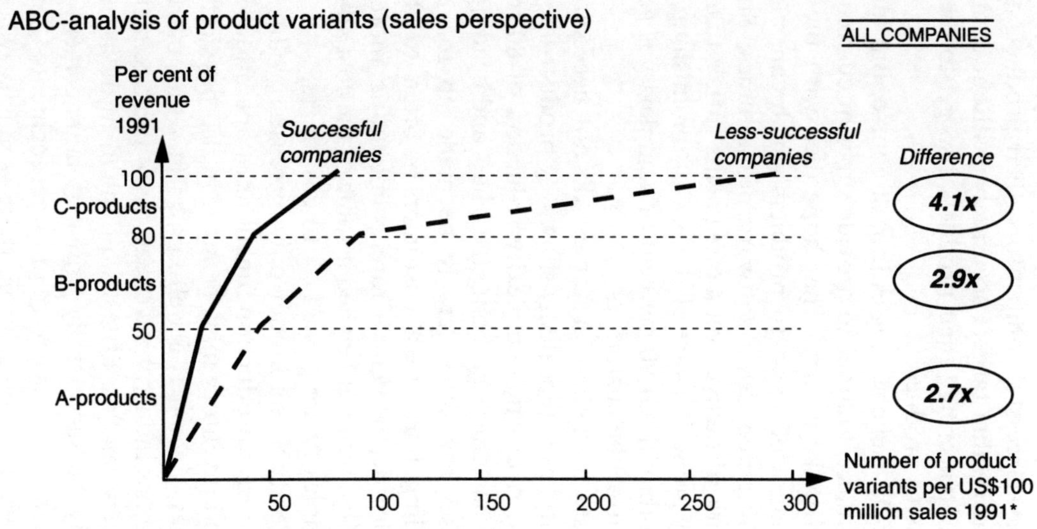

Note: Not comparable to manufacturing – related variants data
* Currency conversion at industry PPP

Fig. 3.3 *Successful companies have four times fewer C-products*

ing simplicity and clear focus. Products that stick to the shelf have no place in their product portfolios, and they have a quarter of the less-successful players' number of C-products (see Figure 3.3). This generally does not turn them into one-product companies, though – their customers, too, expect a certain level of variety. But compared to those of their less-successful competitors, their products are more likely to hit customer-value targets accurately, without compromise but without 'over-engineering'; their products also get to the market earlier and cost less to produce.

There are successful niche strategies too, of course. But increasing development costs make it harder to provide superior customer value for niches and still make money. That hope died with the steadily improving performance of standard products which can, moreover, be more easily configured using software programmes. Successful companies that work attractive volume segments rather than pursuing a niche strategy are well aware of the law of diminishing returns. They have realized that about 90 per cent of complexity is generated for only 10 per cent of the customers.

But in many cases even these customers do not necessarily need such high complexity, as is repeatedly shown by product-line streamlining programmes. A drastic complexity reduction of 80 to 90 per cent will not see all customers grabbing their hats and heading for the door. Volume losses can often largely be made up for with the company's top-selling lines. Lost customers – often to the great surprise of the sales department – rarely amount to more than 3 per cent, and are often well outweighed if some of the cost advantages from reducing complexity are passed on to customers.

Reducing complexity therefore not only increases productivity, in such areas as the indirect production functions of operations scheduling, purchasing, and inventory management. It also pays off directly through higher sales, since it is a well-known fact that markets like those of the electronics industry never say no to price reductions.

The fine line between 'as little as possible' and 'as much as necessary' can only be successfully drawn if management has the courage to be economical in its choice of target segments – and if it is a sound judge of what customer-value these segments will pay for. Successful companies have discovered a simple but effective way to find this out: they gain their customer knowledge first hand. To provide the customer-value identified as fully and with as little overlap as possible, they perfect their product portfolio management, simplify structures and procedures on all levels of the business system, and are very selective about where to make their own in-house contribution along the value chain.

First-Hand Customer Knowledge

How do you design a product so that customers will buy it? When is it time to part with a product or a business and reallocate resources? How radical should the slimming process be? How much complexity is appropriate to new product specifications?

Successful companies decide on all these points mainly with customer requirements in mind. The customer knows what he needs better than the R&D department, or the sales and marketing department. So the device of top companies is: 'Customer contact for product development managers!' or 'Get customers onto the development teams!'

Only the truth straight from the horse's mouth – about the customer, competitors and technological developments – will set a company off in the right direction. Sony and Panasonic, for instance, have been running 'antenna shops' for years in locations heavily frequented by their customers' opinion leaders. They present prototypes of new products in exclusive shopping centres or selected airports. An expert development engineer or product manager is on site, and so receives customer opinions which are then fed back directly to the responsible development teams at the company without major detours or multiple 'translation'.

The type of dialogue companies choose to have with their customers – whether they 'capture' customer opinions in the store or pick them up casually during service visits – is immaterial. What is important is that they receive information direct instead of by hearsay. The old argument frequently produced about R&D staff no longer holds water. ('Our development staff are paid to develop, they're too expensive and there are not enough of them to allow discussions with the customer.')

There are still all too many companies where the need for direct communication without the mediation of interpreters, internal or external, is still not being met. Many still use competitors' products as the only source of information on what is happening in the market. They regularly compare other companies' product features with their own and forget that competitors' product design and satisfying customer needs are two different pairs of shoes. Might it not be possible that the competition, too, does not have the features to respond totally to customer demand? Or is there a chance that competitors have perfect products at the moment, but will soon have to change direction?

Measuring performance against the competition is always a look in the 'rear view mirror', not into the future. Competing products currently on the market are separated from the realization of a company's own plans by at least one and often two product generations. Even though customer needs were adequately met yesterday, they may not be tomorrow. If a company models its behaviour solely on what competitors have to offer it will be an imitator, not an initiator. In the worst case, the company could start catching up on competitors' product features that have no value to the customer

Watching competitive products in the rear-view mirror must, therefore, always be accompanied by looking ahead – at the customer and latent customer wishes. Successful companies regard direct contact with the customer as by far their most important source of ideas.

There is no risk that maximum orientation to customer value will trigger even more complexity. The customers themselves are not bothered about it. They want straightforward product series with easy-to-understand features. Focusing on customer value is the best protection against getting into the complexity trap. And, no less importantly, focusing also helps you get out of it.

To break out of the complexity trap, a company will need all the help it can get. Courageous strokes and bold cuts are its inescapable fate, incremental adaptation is just not enough. Accumulated overhead can only be effectively attacked if complexity is substantially reduced.

The electronics company that achieved an essential turnaround in a major business area thanks to variant management needed to reduce its product portfolio by well over half. This was accompanied by a redesign of its products that cut the number of associated complexity drivers (in this case: SMT sub-assemblies and components) by over 80 per cent. This trimmed unit costs by 40 per cent, leading to an increase in sales of over one-third.

Small adjustments hardly ever bring about the desired success. But just how deep the cuts have to be is almost inconceivable when they are still 'on the drawing board'. Many consider a 30 per cent cut in complexity to be a bold step – only to find later that their overheads have remained (almost) unmoved. Electronics is an industry where cost reductions normally show little impact unless 60 to 80 per cent of subassembly variants are pared away. Such estimates can only be made reliably if a company has a very good understanding of customer needs and circumstances – the kind of understanding that can only be gained by direct contact.

Perfectly Tuned Product Portfolio

In a company where all structures and processes are geared to high complexity, the expensive and complex processes with all their attendant costs will not change if only one or two variants are dropped. And successful companies do not stop at this when managing their product portfolio. They know that certain thresholds can only be passed with radical cuts that make a transition to simpler processes and procedures possible.

One example is reducing the number of components and printed circuit boards to the extent that insertion lines no longer have to be retooled for different boards. Fixed tooling not only eliminates set-up times but – more important still – makes possible much higher process stability, leading to increased levels of quality and the elimination of board testing and reworking (more on this in Chapter 4).

Successful companies have simpler building blocks. They manage to make more finished products out of fewer modules. In the computers/communications segment for instance, less-successful companies have on average 1.7 times as many product variants, nearly four times as many sub-modules, and 4.5 times as many parts. Consequently, successful companies develop 2.3 times fewer sub-modules per product and need to run only half as many different designs through production (see Figure 3.4).

The cost impact of modular design is well understood by today's successful technology-based companies. Apart from enormously simplifying their production processes, modularity also strengthens their position *vis-à-vis* suppliers, opening up another avenue to lower cost and higher productivity. Savings of up to 30 to 40 per cent of the cost volume addressed are quite typical if a high level of commonality among individual modules can be achieved.

Less well-known are the beneficial effects of modular design on the development process and consequently on innovation and expansion rates. Uncoupling core technologies and product features into individual modules and developing them independently with different development cycles allows faster innovation and higher productivity

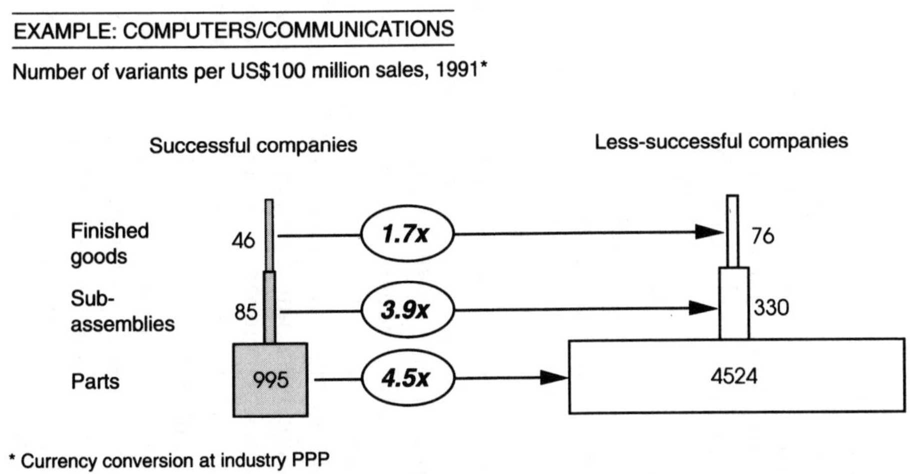

Fig. 3.4 *Fewer variants on each level with modular system*

in the development process, and also enables a company to respond faster and more flexibly to what is happening in the market (more on this in Chapter 5).

Standard libraries of preferred component lists promote the modular principle. The shorter the list and the fewer the unlisted parts used, the better. The best computer manufacturers use parts from the preferred list only, and do all they can to stop the list from 'mushrooming'. An engineer wishing to use a new component has to put up a case for it, show all its advantages and, in exchange, delete at least one old component, in some cases two, from the list. The component may only be used if the need is proven and the deletion is accepted.

In the electronics industry this ideal state is far from universal. On average, 73 per cent of parts used are on the preferred list at successful companies, the figure is only 32 per cent at less-successful companies. Yet the less-successful companies' lists are far longer – making them more difficult and complicated to use.

A further core topic in active portfolio management is how to handle obsolete products. The courage to slim is shown not only by companies who limit their number of variants, but equally by those who think ahead in phasing out their obsolete products and are not afraid to cannibalize their own product portfolios with new product series. Rather than expose themselves to the competition, they take the initiative themselves. Attacks are most likely in the attractive high-volume segment – and that is why companies start revising product lines at the bottom end.

Radical slimming is also a feature of the portfolio management of companies aiming to break out of dead-end niches and try their luck in volume markets. Renouncing their expensive technological extras, they nearly always stop further development of their up-market products. Logically enough, they do not develop their new low-priced standard series by stripping down high-end products, because they do not want to be left with the residual costs inherent in expensive design. They start from scratch, and are well-advised to do so.

Simple Structures and Procedures

Breaking up large, complex structures is the ultimate focusing tool. If a newly-created small unit is dedicated to a specific customer or product segment and given entrepreneurial freedom – that is, sufficient decision-making powers and independence from the parent organization – complexity goes down automatically. Focused pipelines for product development, where it is the products and customer needs – not the technology – that are central, are a proven tool for optimizing development work. Each product line is looked after by a small, empowered unit that coordinates customer wishes, development interests and manufacturing requirements. It concentrates on its product and takes full business responsibility for it. Functional conflicts of interest and associated friction losses are thus prevented from the start. Businesses rather than functions are optimized.

In manufacturing, too, these focused pipelines can be used as a basis for high productivity. The aim is to have hit products coming off the assembly line without interruption, without much retooling, in fact preferably with fixed tooling. Hugely popular and exclusively appreciated products are made on separate lines, manufacturing is segmented by product lines and not type of production. For example, an injection-moulding machine for plastic housings may easily be integrated with a printed-circuit-board assembly station.

With increasingly advanced production technology, the unit numbers needed for full utilization of a production line are shrinking constantly. If it cannot fill capacity despite this fact, a company will aim to form clusters of subassemblies using the same range of components. This concept is ideal for the manufacture of printed circuit boards. It allows full capacity utilization and fixed tooling as well. The range of components in the automatic insertion equipment need not be changed, machines are available longer, and process stability is improved.

The Rosemount pressure transmitter factory in Chanhassen, Minnesota owes its success to a new manufacturing concept of this kind, highlighting the complexity-management issue. Faced with a large number of measuring ranges, different materials, and hundreds of thousands of combinations, it introduced a modular design – a few standard modules for all transmitters. These standard modules are then put together on dedicated final assembly lines. High product variety – and therefore manufacturing complexity – is not generated until the last stages, a long way downstream in the production process.

An intelligent arrangement of the lines in final assembly also helps employees to adapt quickly to the product variant being manufactured. U-shaped lines shorten communication paths. Staff can respond immediately to alterations in the production process and in materials. At Rosemount, this production strategy caused a revolution: between 1990 and 1993, manufacturing costs were halved and throughput times fell by 62 per cent. At the same time, process quality rose significantly.

As the Rosemount example shows, modular concepts mean that products can be tailored to customer needs in the final stage of production. The complexity of the product range does not complicate the entire manufacturing process, and individual customized solutions are produced right at the end using a small number of modules. PC-maker VOBIS takes much the same approach – but in an even more extreme form. The customer's desired configuration is assembled not in the factory, but at the retailer.

Companies that put their faith in volume business and the building-block principle spend far less on automation. With fixed tooling or limited retooling, an expensive and flexible plant would be a pure luxury. Between 1992 and 1994, successful companies invested less than a third in automation compared to less-successful players (as shown in Figure 3.5).

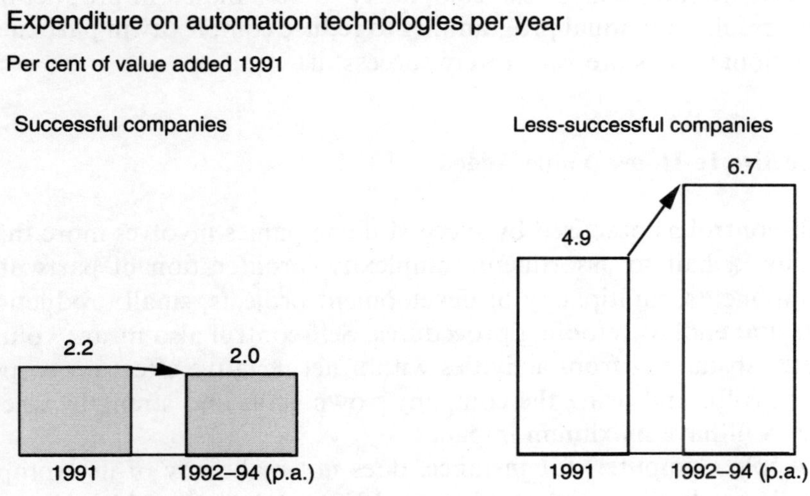

Fig. 3.5 *Investments in automation do not guarantee success*

Foxboro, a US manufacturer of industrial electronics, has used uncoupled structures effectively to tackle the complexity generated by old products. Its solution – rigorous separation of old and new product manufacturing – is an extension of dedicated production. Labour-intensive manufacturing of replacement subassemblies for the control systems that are installed in almost every nuclear power station in the US has been transferred to the company's subsidiary in low-cost Shanghai. The new products – untrammelled by the complexity of the old – are churned out smartly at highly-efficient US production facilities.

Unnecessary complexity in manufacturing can also be avoided by consistently exploiting technological progress. Things have changed, and the same hardware can now be used for many different purposes; the software is used for tailoring to specific situations. This also applies to the input and output subassemblies of complex control systems. Where 45 different subassemblies were once needed, one subassembly for digital connections and another for analogue connections now do the same job.

The necessary adjustments for the measuring ranges and bus systems are made with the software. Many subassemblies therefore no longer have to be made. The example also shows how important development is to manufacturing. The product development process lays the foundations of the complexity to be handled in production. As a result, individual programmes to reduce complexity in particular functional areas are very rarely successful.

Selective In-House Value-Added

Self-control as practised by successful companies involves more than calling a halt to assortment complexity, proliferation of parts and components, multiplicity of development projects, small production lots and endless retooling procedures. Self-control also means voluntarily abstaining from activities with a net negative effect on corporate profit, and using the company's own skills and strengths where they will have maximum impact.

Apple Computer, for instance, does not make any of its laptops, the Power Book or the new Personal Digital Assistant (Newton), in-house. Apple sees its principal contribution as being in development, and that is where the company has focused, leaving manufacturing to Sony, a company with high manufacturing skills.

The extent to which a company chooses to integrate its value chain depends ultimately on its strengths and weaknesses. A company with superior technological and operational strengths will tend to be relatively highly integrated, since integration gives it a competitive advantage. A company with weaknesses in these areas will naturally tend to outsource. Companies aiming for the top from a position of weakness should reduce their own contribution, as long as they retain the innovative value-adding stages in-house (see Figure 3.6).

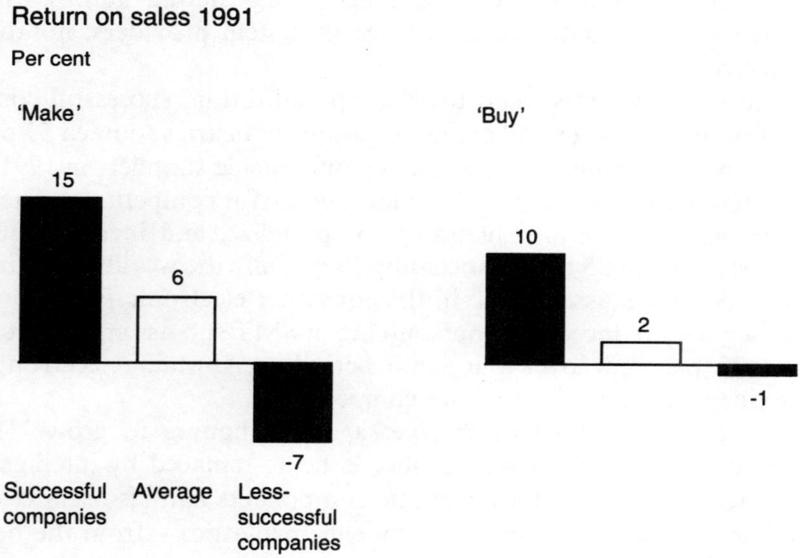

Fig. 3.6 *Higher leverage for in-house production*

The make or buy issue will concern the electronics companies of the future even more than it does now. The only defence against the increasing speed of innovation and rising cost pressures will lie in winning a large share of the world market. This means paring down output to a few products, components and modules. Some products and modules will take the company into the league of global suppliers, others will be discontinued and outsourced completely. At that point, the key factors for success will no longer be a large total market share and control of a wide span of the value chain, but mastery of key technologies and a large world market share in certain key components or modules.

Companies have already begun concentrating on particular modules and value-added stages and, with increasing product modularity and standardization of interfaces between products and modules, this trend will grow. This is particularly evident in the computer and communications industries and in consumer electronics. Disk drives, hard-disk storage and motherboards are supplied almost entirely by specialists. In consumer electronics, for example, Sony and Philips are the world suppliers of laser modules and mechanisms for CD players. Cathode ray tubes, too, are supplied by only a handful of manufacturers. Similar trends are appearing among automotive-industry suppliers with the emergence of system providers, notably for electronics.

In keeping with this trend towards specialization, successful companies in the computer and communications industries sourced 52 per cent of their electronic subassemblies from outside suppliers in 1991 – more than twice as much as their less-successful competitors. These poorer players were also planning to specialize, and increase their outsourcing from 25 to 35 per cent by 1994. Only the world's best and most productive 'assemblers' in the consumer electronics/small products segment reduced their outsourcing of SMT sub-assemblies from 33 to 27 per cent over the same period. Assembling electronics components is part of their core competence.

Focused value-added means, yet again, trimming to grow. The corporate philosophy of self-reliance is being replaced by intelligent outsourcing of the most appropriate components and also services – whether in manufacturing, development or logistics – from the best supplier.

SUMMARY

If a product contains too many features, particularly ones that the customer is not prepared to pay for, unit-cost disadvantages arise that cannot be compensated for or – better still – outweighed by price premiums for extra value. Successful electronics companies systematically avoid the disadvantages of overcomplexity and over-long internal value chains. They allow only as much complexity as is needed to satisfy customer needs well in their target segment. They know that focusing boosts earnings and creates new growth.

- Sustainable corporate success means slimming down to the most important target segments. The product range should reflect customer segmentation as clearly as possible and customer segment products must be tailored to meet their exact requirements – the essential understanding of the customers, however, can only be gained by direct contact.

- High-end niche products often come under pressure from the volume segment. They then become profit killers, adversely affect the cost position, and obstruct the view of attractive volume markets.
- Complexity costs of all kinds that have increased incrementally cannot be reduced incrementally; radical cuts are needed.
- Only by using refocusing as an integral part of corporate activities will long-term growth be achieved and profit be sustainable – particularly in the price-sensitive electronics market.

4 High Costs are Mostly Home-Made

Successful electronics companies owe two-thirds of their higher return on sales to lower manufacturing costs. Manufacturing, conversely, is also where less-successful companies give away profits. Does this put world-class manufacturing beyond their reach? Are European companies already irretrievably far behind? The answer from the long-term study is 'no, but . . .' No, because the most important levers can be influenced within the company itself. But, because radical measures will have to be taken to close the huge gap and then achieve the improvement rates of the world's best players.

Thirteen cents per dollar earned is a lot of money, especially in an industry with narrowing margins. And this is the amount by which the best electronics companies outperform their weaker competitors, on the production cost for their entire product ranges alone (Figure 4.1).

The extent of the gap in manufacturing unit costs between average European suppliers and world-class companies is revealed by reverse engineering. Differences of 40 to 60 per cent for individual, directly comparable products are not uncommon. Mention findings like these, and the response will normally be arguments on the choice of location and stories about unbeatably disciplined workers in the developing and newly-industrializing countries of Asia. Such arguments are, however, beside the point. Worse still, they distract attention from the fact that the ability to make the most important improvements already lies in management's hands.

The starting point for these improvements is revealed by a close analysis of the production cost gap. Differences in design account for at least half of this gap, with operational inefficiencies and factor cost differences – in about equal proportions – making up the other half (see Figure 4.2). There are thus three main levers to improve manufacturing productivity:

- *Design for manufacture* must be added to design to market: the attractive product derived from focusing on high customer value must also be easy and inexpensive to make, install and service.

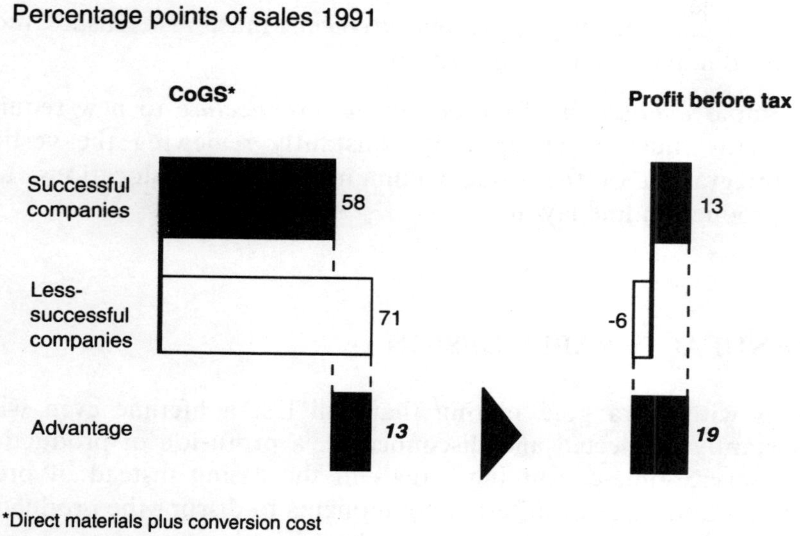

Fig. 4.1 *Better ROS at successful companies is largely due to low manufacturing costs*

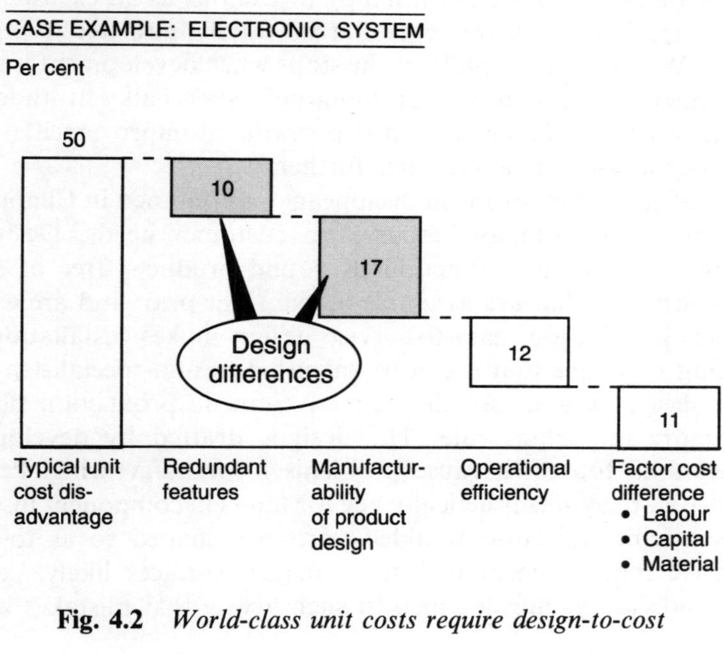

Fig. 4.2 *World-class unit costs require design-to-cost*

- Fast *operational process improvements* must also ensure more productivity, quality and speed.
- Rapid *adaptation of the production architecture* to new requirements must be ensured by constantly reviewing the vertical integration of the value chain, manufacturing locations, and production line layout.

MANUFACTURABLE DESIGN

Plugs with extra gold plating that will last a lifetime even when constantly connected and disconnected, a profusion of production- and service-time-consuming screws in the casing instead of press-studs, a multitude of different components to disrupt the production flow and fill the warehouse, no self-diagnosis systems to simplify repairs – the list of design faults could be continued indefinitely.

Such failings immediately push up costs. But a further danger, often overlooked, is the negative impact they have on motivation. When employees see that problems in manufacturing, installation and servicing (or in operation at the end-user, for that matter) are due to poor design, they are often happy to use this as an excuse not to make operational improvements that are totally unrelated to product design. 'Why should we pull out the stops when development is giving us products that just hold everything up?' Such balky attitudes can lose the company urgently needed operational improvements – and its unit-cost position deteriorates further.

Part of the design problem disappears – as outlined in Chapter 3 – as soon as the company focuses on customer needs. Design-to-customer then ensures functionally sound products, free of superfluous features, that are available at the right price and are easy to operate. In addition, easy-to-service design makes installation and servicing so simple that it can be entrusted to non-specialists.

But design weaknesses that directly impede production play an even more important role. The designs drafted by development engineers all too often cause problems in manufacturing. Printed-circuit-board layouts practically beg for faults in component insertion and soldering; injection-moulded parts are shaped so as to make expensive moulds necessary and damaged surfaces likely; components and casings are designed in such a way that mistakes will be made unless complicated assembly instructions are followed. Yet electronics manufacturers have found that good design alone can reduce defect rates by 50 per cent.

Manufacturability of design means that the product's engineering and layout alone ensure high process reliability. Such intelligent products, taking account of manufacturing requirements, are still far too rare. The goal should be to produce a design in which fluctuations in the quality of materials and mechanical and human 'intervention' have as little effect as possible on the quality of the end product. High quality also requires high process stability, and an important lever for achieving this is a high share of identical parts, resulting in large production lots.

Intelligent Design

The fact that process robustness fails more often than necessary is hardly surprising. And it will continue to do so as long as development and design engineers think up their solutions in splendid isolation at their desks without feedback from production-line workers and without adequate computer support for design work, information management and simulations. How can a development engineer be expected to know – and keep bearing in mind – all the tiny details that can disrupt production and frequently hold the potential for relatively large savings? How is a development engineer to know what is really 'easy' for manufacturing and assembly, without detailed knowledge of the procedures, machinery and people in the factory?

Fast improvers and highly productive companies accordingly build in close contact between development and the production line, and – led by the Americans – ensure that their design departments receive the best possible systems support, with module libraries and simulation environments for both software and hardware. Their CAD systems include standard design elements, for example, and they also keep on file a detailed knowledge of some specialized subjects, such as the design of surfaces and corners for injection-moulded parts or the positioning of components on printed circuit boards for optimum insertion and soldering.

Leading manufacturers are also making increased use of simulations to detect sources of faults in product and process design preemptively, using the FMEA method ('Failure Mode and Effect Analysis')* to make a systematic analysis of the production process from the point of view of potential sources of error in the product or the system. This is based around experimentation with the idea of 'What will happen if this goes wrong?' The entire manufacturing process unfolds step by step, the effects are documented, and action is taken to eliminate or contain possible errors in the product or the process.

In addition, the Taguchi method† can be applied – with relatively few experimental and analytical steps – to achieve the optimum fit of product characteristics and process parameters in complicated production processes. In simple terms, this approach is a scientific/experimental complement to the FMEA method. It facilitates the development of a 'robust' design specifically for complex products or processes in which a large number of parameters have to be optimized simultaneously.

For example, a quality function may be generated for processes on the basis of the relevant parameters (for example, the number of non-conducting soldering points as a function of soldering temperature, speed, component shape, or pre-treatment of the contact points), and a highly-efficient experiment plan drawn up to analyze the relationship between the ultimate product or performance characteristic and the parameter settings to be optimized.

Alongside these 'intellectual' tools and methods, computer-aided simulations are also becoming increasingly popular. Successful electronics companies use them to simulate not only the manufacturing process, for example, printed-circuit-board assembly and final assembly, but also product functionality under disruptive ambient influences, such as the reaction of circuits to electromagnetic fields. Such simulations obviate the need for long, expensive revision loops, and the first – or at the very latest the second – draft design can often be implemented with no further revision.

---

* Gilchrist, F., 'Modelling Failure Mode and Effects Analysis (FMEA)', *International Journal of Quality and Reliability Management*, vol. 10, no. 5, pp. 16–23.

† For an introduction to Taguchi methods, see Ealey, Lance A., *Quality by Design*, 2nd edn, ASI Press/Irwin Professional Publishing, Burr Ridge, Illinois, 1994.

Finally, successful electronics companies refine their knowledge of process-oriented design engineering in the course of development and pilot-series manufacturing. This knowledge is then stored in handbooks, expert systems or computer-aided libraries for ease of access, and is also updated regularly.

Stable Processes

However intelligently individual products may be designed, they may still fail to meet the 'robust processes' standard. If the product is allowed through to the manufacturing stage with too many new and different elements, manufacturing will be overburdened. Errors will proliferate. Successful companies therefore stick to three principles that, once again, have a lot to do with focus and simplicity.

- First, they work with only a *small number of different technologies*. They use few elements with different shapes in their electronic subassemblies; they minimize the machine operations for metal and plastic parts.
- Second, they *re-use proven modules*, not only for the products in a single product family, but also for next-generation products. Open up a number of Panasonic telephones, for example, and you will see that the same circuitry has been used across several different products and product families.
- Third, they *approach new technologies with due caution*. They do not retire an established technology until it is proven that a new one will improve functionality or increase productivity. The transition is carefully evaluated up front. But once the decision is taken, it is implemented fast. Top companies are uncompromising in their implementation of innovations – witness the replacement of mechanical and electromechanical components with electronic ones, where they took a much more targeted and systematic approach than slower competitors.

Successful companies also lead the way in the transition to new surface-mounting techniques for electronic components. They changed earlier on average to the new SMD (surface-mounted-device) technology from the traditional PTH (pin-through-hole) technology or manual assembly (see Figure 4.3).

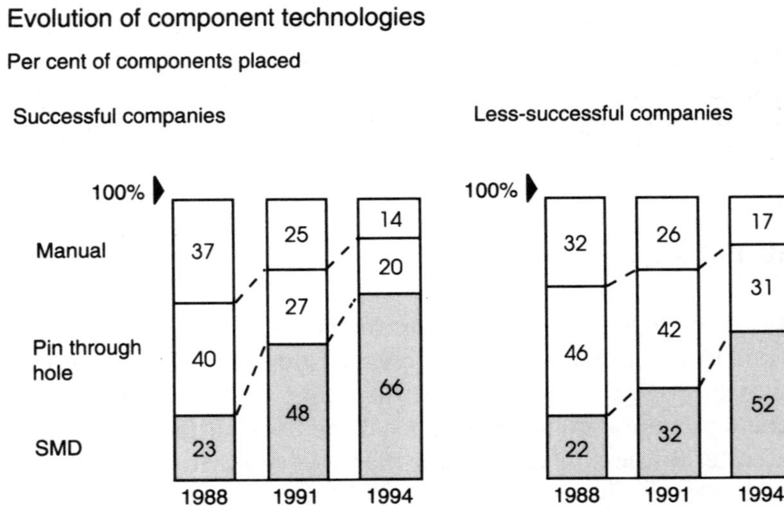

Fig. 4.3 *Faster transition to surface-mount-device technology*

A close inspection of the PC boards in products from successful players confirms the pragmatic and highly-targeted combination of old and new: State-of-the-art technologies for chip assembly on printed circuit boards and miniscule SMD units are used wherever justified by customer value. Palm-sized video cameras and 'Handy' format mobile phones require miniaturization and the tightest possible integration for reasons of space.

These products show that successful companies learned how to handle the technologies needed here in good time. Products made by less-successful suppliers are often conspicuous by their bulk, and a glance inside shows that the manufacturer is still struggling with production and component technologies that lag two generations behind world class.

The appearance and technical standard of the world's best products are quite different where huge size restrictions or high innovative pressure are immaterial. This is the case, for example, in television sets, where surprisingly old circuit modules and components can often be found. And as factory visits show, these sets are made on equally old, dedicated production plant, often even in specialized factories in low-wage countries.

FAST OPERATIONAL IMPROVEMENTS

The further development of the manufacturing processes goes far beyond the 'complete redesign' of products and processes. Successful companies pay undiminished attention even to seemingly small improvements, and keep on making them. The results of these continuous improvement programmes are amassed into a treasury of experience that is constantly fed into new product generations and new manufacturing processes.

The impact can be observed in automation practices, for example, where successful companies have a very much better sense of proportion. And here again, the Japanese provide particularly striking examples of the success of continuous small improvements.

Automation in Proportion

Less-successful companies automate too much, believing that ultra-modern, fully-automated, and closely-interlocking plant meets the needs of the dynamic electronics markets best. They are trying to achieve the much-cited 'Lot size of one' through fully-automated tooling. However, unless product and process design have been sufficiently focused and complexity has been greatly reduced in advance, such an approach leads to over-expensive, unproductive machinery (see also Chapter 3).

Companies with automation overkill generally think machinery is more important than increasing labour productivity. And this is where they make their next big mistake. Because they only count direct wage costs, currently amounting to between 2 and 5 per cent of production costs, they expect this area to have little or no impact on total cost. They overlook the far greater share of indirect costs and the interrelationships with the costs of premises, machinery and materials that do make an increase in labour productivity very much worthwhile.

The difference in production costs mentioned at the beginning of this chapter mostly lies in conversion costs – the costs of transforming purchased material into finished products. In turn, labour productivity plays a crucial role in conversion costs. In the computers/ communications segment, for example, successful companies require just 3.5 employee hours in production to deliver US$1000 of value-added, whereas less-successful ones need 10.3 hours. There are equally striking differences in large systems, where the figure is 3.3 hours against 7.3 hours for those same US$1000. In consumer electronics the gap is smaller, at 20.9 hours versus 24.4 (see Figure 4.4).

This means that, on average, successful companies require about half as many employees in production (direct and indirect) as their competitors in these fiercely contested segments. Such a large lead is, of course, not to be lightly disregarded. For one thing, it gives a company clear operational advantages. For another, investments in machinery and automation will be viewed in a quite different light in factories with extremely high labour productivity. Whether it is people who are to be replaced or productivity that is to be further increased by machines, the standards set for the need for such investments will automatically be much more demanding.

By contrast, once they have been bitten by the automation bug, less-successful companies will soon start keeping a lookout for the next problem to be solved through automation. Their investment in automation is consequently more than double that of successful companies (see also Chapter 3). For instance, such companies often try to make up for a low component insertion rate by changing over early to more highly-integrated switching circuits and reducing the number of components. This certainly reduces their cost of insertion, but the cost per component increases because the chips required have to be specially designed and produced in relatively small lots. On top of this, they also have to absorb the acquisition costs of new, expensive automatic insertion machines and bear the risks of developing and producing new chips – none of which brings them any closer to the productivity rates of fast companies in the US and Japan.

Successful companies take a different route. They not only exploit the full productivity potential of their selected approach from the start with design for manufacture, but also continuously improve the functionality of existing machinery, thereby achieving higher process reliability with fewer breakdowns and smaller capacity requirements. As a result, their automatic machinery has a longer useful life, reducing their investment needs and changeover risk.

Companies with high process reliability also achieve better quality. This saves on testing and reworking, freeing up labour capacities for other tasks. For example, if the insertion heads of an automatic PC

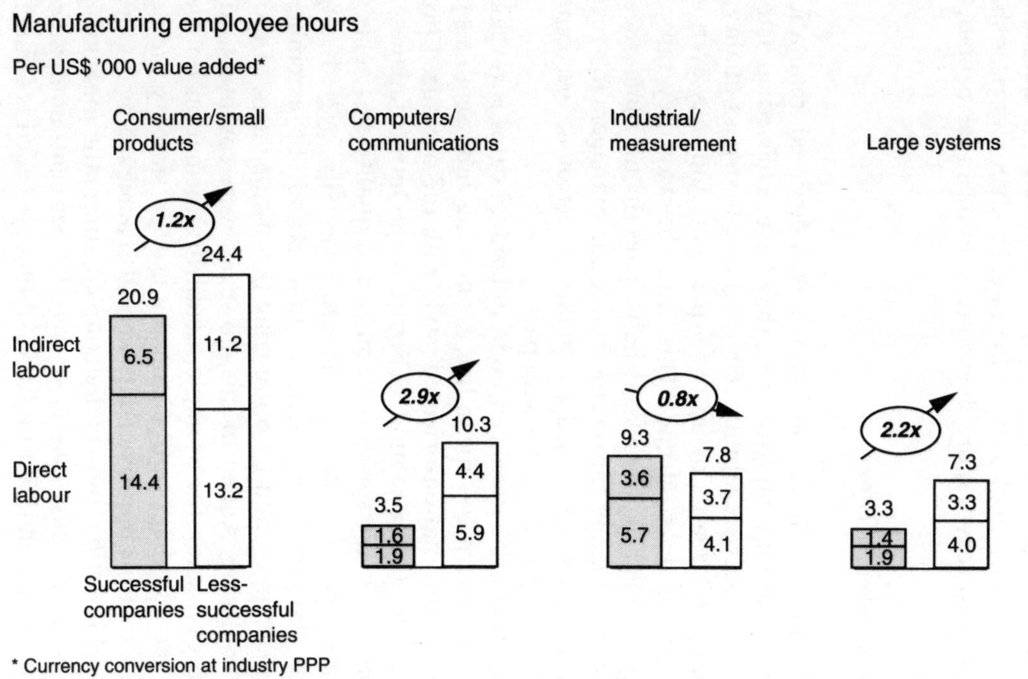

Fig. 4.4 *Fewer manufacturing personnel*

board assembly machine are set – at the suggestion of the workers – so that fewer pins are damaged and the automatic inserter can run uninterrupted for a whole hour instead of a few minutes, production line employees can be solving other problems, can operate more machines, and ensure timely supply of materials, minimizing the stoppages caused by stockouts on the line. Productivity rises sharply and the company leaves more dilettante optimizers far behind.

The Japanese Success Story

Japanese companies have long had a well-deserved reputation as masters of continuous improvement; they have achieved an astonishing degree of process reliability. They achieve first-pass quality rates in assembling PC boards of up to 99 per cent, and may even attain 100 per cent for days on end. This, and the way they talk of an availability problem if a stockout halts a production line once in a year, confounded many European production managers for years. In Europe, companies were used to quality levels of 85 per cent and weekly down-times due to missing parts.

How high the differences can be in individual cases is revealed by comparing a western European line for making subassemblies in the conventional 'automation-oriented' style with the optimized production lines of the Japanese model (Figure 4.5). On the Japanese-style production line, higher process stability and quality was achieved in the first instance by product redesign, especially by reducing the number of 'exotics' – components that cannot be assembled by standard machines – and by optimizing the arrangement of modules for easier soldering. Added to this, however, were continuous operational improvements; for example, detailed information on the fine tuning of the automatic equipment was used to simplify the flow of materials. Extensive improvements were also achieved by doing away with manual insertion and rigidly-coupled automatic machinery.

By boosting first-pass quality to over 99 per cent, companies can do without an in-circuit test on line. All that is needed is a functional test after assembly and an automatic test of the inserted components using a component sensor. As a result, the production line needs only four workers instead of 20. Design optimization and the switch to fixed tooling means that it is not necessary to have several automatic inserters of the same type. Line capacity is almost doubled, and investment accordingly halved. In parallel to this, much less space is

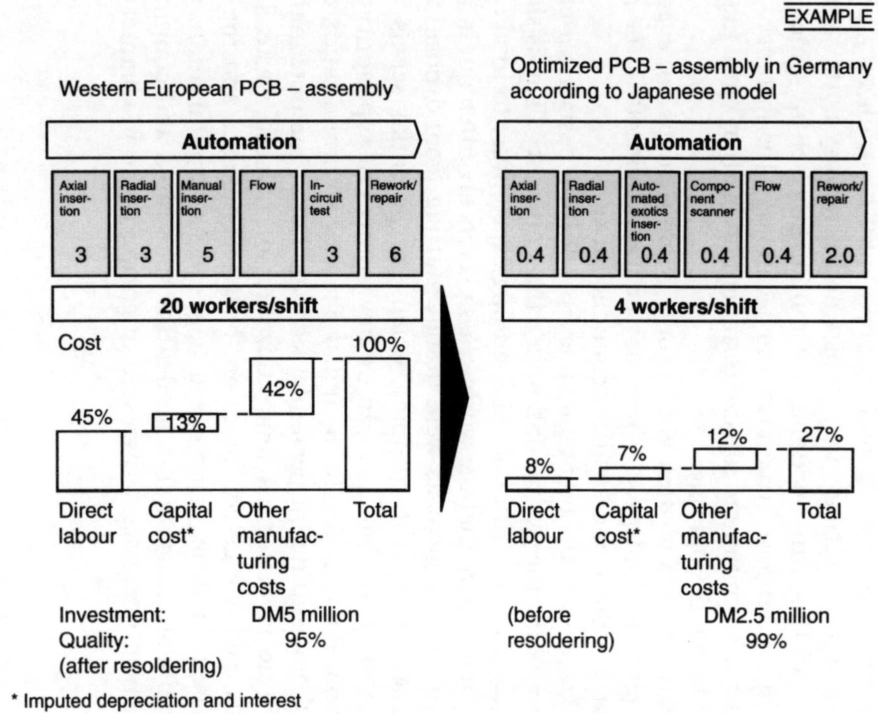

Fig. 4.5 *Drastic improvements also possible in Germany*

needed and other manufacturing costs can be reduced by a good quarter.

The total cost saving is 73 per cent, accompanied by higher quality, delivery-reliability and capacity.

It still comes as a surprise to many people that improvements are still possible even after several waves of optimization. But successful process optimizers keep finding ways to reach higher levels of performance. They set themselves apparently contradictory objectives – such as 'lowering tooling costs and inventory costs simultaneously'. If this can be done, by minimizing set-up times and simplifying warehouse structures and technology for instance, they arrive at a new 'optimum' lot-size formula that is often more than 50 per cent below the old one.

But halving costs is often not the only – and perhaps not even the most important – outcome. There may be useful spinoffs in the form of higher flexibility and smaller economic lot sizes.

In practice, even the best companies' production costs are a factor of two above the theoretical optimum – the 'core costs'. This is shown by an analysis of the ratio of value-adding expenditure to total costs.

No wonder, then, that the world-class players are often out in front in rate of improvement as well, despite starting from a position of advantage. Instead of stopping when the potential seems fully exploited, they continue to open up new sources of superiority with new action programmes. In the initial phase of an optimization cycle they will concentrate on personnel savings. Later, the potential will normally lie in better machine utilization or reduced waste, input materials, and energy consumption. New technologies and the products based on them then start a new cycle of optimization. (The organizational infrastructure needed for such rapid and continuous improvement processes is discussed in greater detail in Chapter 8.)

RAPID ADAPTATION OF THE PRODUCTION 'ARCHITECTURE'

Siemens reduces its capacity and PKI moves its telephone production into its administrative building. Honeywell clears its old production shops and makes domestic utility controls for large frame air-conditioners in its office building. Japanese electronics manufacturers suddenly close down sites in Japan and move their production to

China. A few years ago, headlines like these would have met with shock and disbelief. Now, such reports seem almost familiar. But in every single case it is still interesting to ask how the development happened, and what the trend means to the future of the electronics industry.

The background to these changes is often similar. Production strategy is reviewed in the light of high production costs coupled with declining production volumes due to rapid innovation and increased productivity. As a result, companies recast their ideas on where to produce, how much to produce in-house, and how much capacity to retain. Successful companies don't just stop short after a slimming diet of design for manufacture, modularity, continuous operational improvement, and changes in the organizational structure and processes. They use the restructuring and redesign of their production 'architecture' as a tool for improving performance.

To this effect, successful companies take steps to concentrate their manufacturing facilities – which are often for historical reasons located in numerous, widely-scattered plants – in order to achieve critical mass in know-how and manufacturing volume at one or a few locations. Aggressive competitors target their choice of location as a way of getting closer to suppliers and markets and/or profiting from low factor costs.

Concentration and Capacity Adjustment

Production locations and plant are no longer designed for eternity. Planning is too uncertain for that in the dynamic electronics industry. Successful companies have realized this, and are adapting manufacturing capacities to production volumes faster and more radically. Compared to their less-successful competitors, they have up to 26 percentage points better capacity utilization (see Figure 4.6), and their profits are higher too.

Such capacity changes cannot be avoided in the long run, because of the way the industry dynamics continuously slash value-added. This is evident from a comparison of value-added as a percentage of input material cost (as shown in Figure 4.7). In the still partly mechanically-based field of industrial electronics, value-added amounts to 334 per cent of the value of the material; in the mature and highly-competitive field of consumer electronics, value-added has shrunk to 61 per cent of direct material cost.

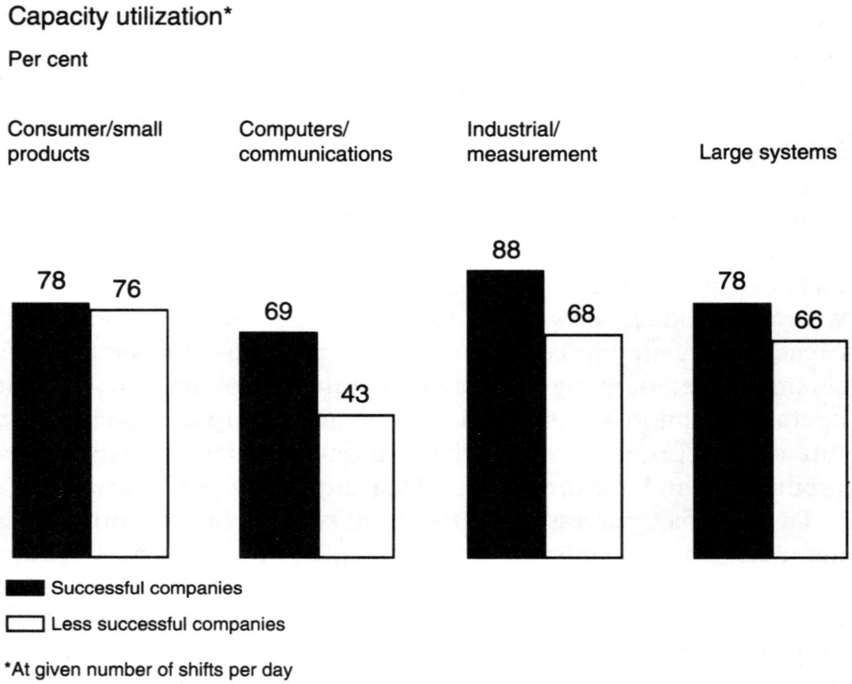

Fig. 4.6 *Significantly higher capacity utilization*

Three main developments are responsible for these changes:

- First, the increasing technical integration of electronics hardware means fewer components;
- Second, design and operational improvements reduce defect rates, set-up times, and other non-productive elements;
- Third, competitive pressure is leading to the rise of component specialists that often turn OEMs into pure assembly plants for modules bought from suppliers or manufactured by one of their own subsidiaries.

In response to this unstoppable trend, successful companies are concentrating similar products at dedicated sites, bundling activities, and closing locations that lack critical mass. At the same time they are adapting machine capacity and manpower to current expectations regarding the changes in value-added, so that they continue to

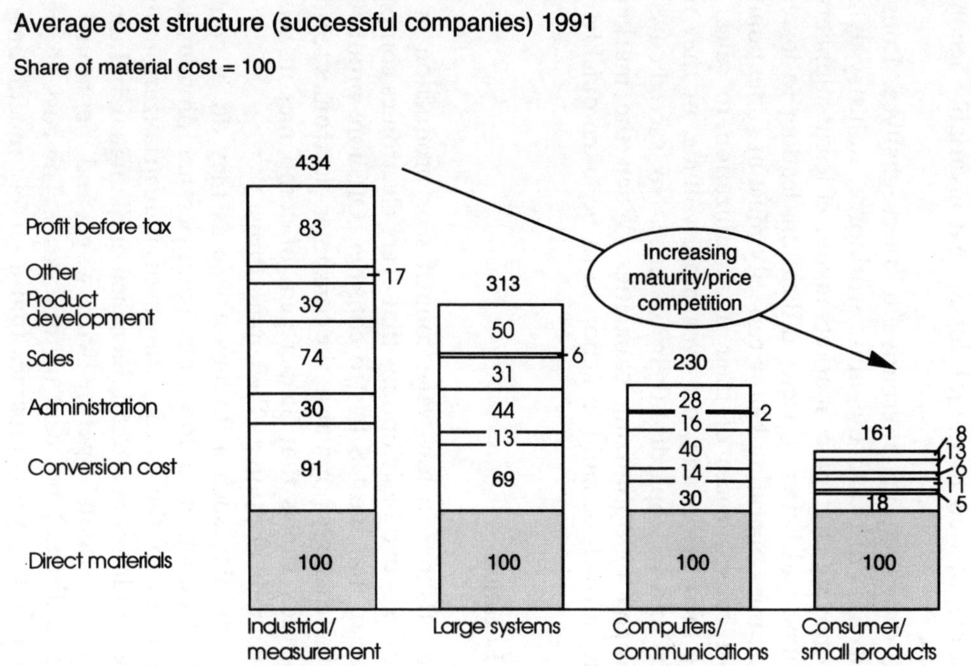

Fig. 4.7 *Value-added and margins decline as segments mature*

achieve their winning productivity increase of approximately 7 per cent a year.

Expanding production sites and opening new ones are the exception in this type of restructuring, apart from transfers of production to low-wage countries. At the same time, facilities that are downsized or closed – a far more likely occurrence – have lost none of their dramatic force, in view of the job losses that invariably accompany such actions.

There is, however, no alternative to these changes. Long-term overcapacity is a threat to the entire workforce. Industry dynamics in the fast-moving electronics markets favour efficient suppliers with higher volume and better capacity utilization, leading to lower unit costs. Companies operating below capacity must at some point pass on the cost consequences of their underutilization to sales. They either raise their prices and lose even more volume, or they reduce their profits and so jeopardize their existence. No 'sacred cows' may be worshipped in optimizing capacities. When the market and productivity growth demand it, action must be taken quickly.

Choice of Location

Production offshore in low-wage countries or production in high-wage countries – two basic options that every electronics company in Europe, Japan and the US has to consider. Offshore production is becoming increasingly popular in the consumer-electronics sector; the long-term study shows that 35 per cent of companies in this area already favour this option as their main thrust.

Where costs play such a dominant role as they do in consumer electronics, production in low-wage countries has obvious advantages. Factories in the border zones of industrialization, in the Chinese free-trade zones around Shenzhen or Shanghai, in Indonesia, Thailand or Vietnam, operate with wage and personnel costs amounting to about 1 per cent of total costs. The cost of premises is also much lower, yet companies find a largely functioning infrastructure with power, water and telecommunications connections already in place. Local workers meet high educational standards that in some cases match Western standards.

These obvious advantages are generally counterbalanced by higher transport costs, lower labour productivity, and language and cultural barriers that make integration difficult. The political and social risks in these breathtakingly dynamic regions cannot be underestimated either.

On balance, the move is still worthwhile for many companies. Sometimes they have no other choice, as when products with a low innovation rate made in stable mass production facilities are subjected to extreme price pressure – particularly when the products are relatively low volume. In terms of landed cost, that is, the total cost of the products in the country where they are sold, a Walkman personal stereo is better suited to offshore production than a television set, since freight costs should not be the predominant part of the cost structure. Consequently, Sony, for example, produces television sets in several regions, whereas Walkman or Minidisc player production is centralized in the Far East.

The most successful consumer electronics manufacturers make skilful use of even the smallest factor cost advantages. Their management has the skill of designing factories so that they can be built and commissioned fast. Within six months, these factories reach almost Japanese levels of productivity. After three to five years, when the rise in wage costs – typical in the newly-industrializing countries – can no longer be cancelled out by productivity increases and the advantageous 'border zone status' fades, these companies pass on like nomads to a new location – Singapore not long ago, Vietnam or Shenzhen today, and perhaps other emerging countries in a few years' time, or one of the CIS countries (see Figure 4.8).

But the game of 'nomadic factories' on the geographic frontiers of the industrially-developing world is only for players who are complete masters of the art. As one production director aptly put it: 'If you can't make your factory productive at home, then you won't be able to do it in Singapore or China either.' Instead of the expected advantages, further setbacks occur.

Any change in location made on the grounds of cost must therefore be particularly well-prepared. First of all, the product design has to be optimized, then the location issue has to be explored with expert support on site, and a competent partner selected to take on local management and operations. In parallel, the home location should exploit every last ounce of efficiency-improvement potential in production and understand exactly which operational performance breakpoints have to be reached in the generally less automated manufacturing facilities of the low-wage countries.

These considerations, already part of everyday life for most consumer-electronics companies, are not as current in other segments of

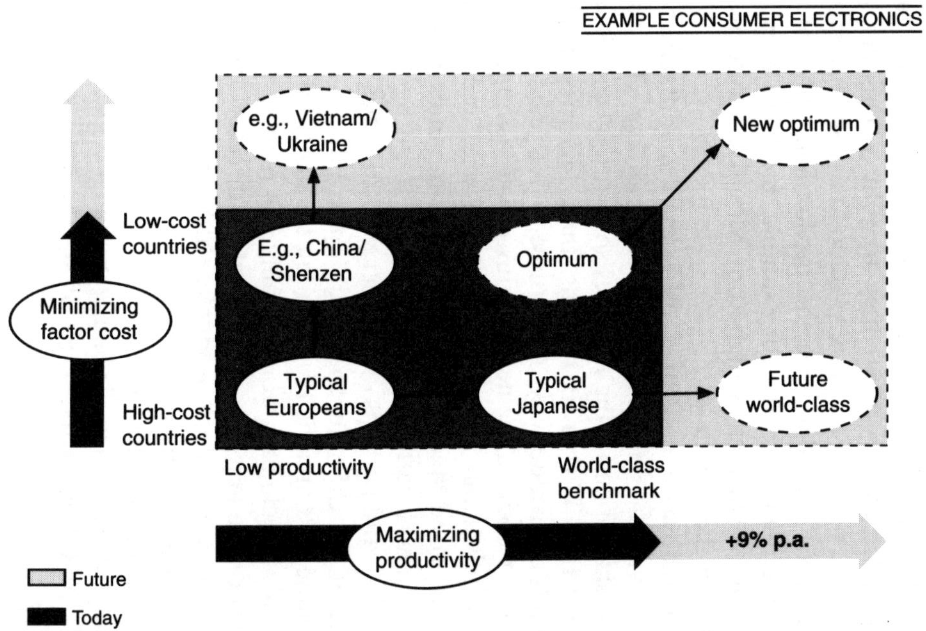

Fig. 4.8 *Improved world-class, new low-cost countries and new total optimum goals threaten optimum goals*

the industry. Here, high-wage countries can still match the factor cost advantages of nomadic factories through a spirit of innovation and greater design and marketing flair, combined with highly-efficient manufacturing. Companies that have followed the general migration to the Far East because it is the flavour of the month, without undertaking the necessary research and preparations, have sometimes made this discovery. They now know that they should keep a significant share of their added value at home.

Considerations similar to those governing the 'nomadic factories' also apply in software development, where even today, linked by satellite, hundreds of software-development engineers sit in Russia or India and work for worldwide conglomerates, transmitting completed lines of code back to head office every day. Careful thought must be given to the division of labour, though. Russian development engineers, who previously had to solve complex problems with less-powerful computers than their Western colleagues, are known for their highly-efficient and effective algorithms. But they still have a lot to learn about using the latest programming languages and tools.

Of course, cost is not the only reason for shifting individual functions around the world. Companies often decide on a new location to get closer to their customers and information markets, or even to their subcontractors.

A presence in many Asian countries, for example, offers admission to the local markets with their enormous growth potential; the local content restrictions imposed by the governments of these countries are not the only entry barriers they put up. Companies are sometimes shut out by their own actions; European companies in particular readily forget that local presence is essential in some businesses. But why should the considerations which led companies to establish individual functions, such as R&D and assembly, in customer-specific markets in European countries and the US, not apply to Asia as well?

And finally, apart from gaining market access, many companies use their corporate presence in the host country to secure an 'insider' position that links them directly to the information channels of local manufacturers and suppliers. In highly innovative areas, for example, companies are often concerned about being geographically close to suppliers and component developers. This gives them accurate and timely information about new developments; perhaps they may also get prototypes for testing ahead of the competition. Or they may be under heavy cost pressure, and want access to the spot markets for standard electronic components that have grown up, for example in Singapore, with prices up to 30 per cent lower than the official list prices in the US or Europe. Access to these (often informal) regional markets is available, however, only to 'insider' companies operating on-site with local employees in the region concerned.

It is interesting to note that the host countries have started a movement away from the status of mere 'transit camps' for nomadic companies wishing to exploit factor cost differences. The governments of some former border zone countries, such as Taiwan and Malaysia, are now introducing targeted development programmes to build long-term special know-how in such areas as chip development and production. They are attempting to find an alternative to their deteriorating factor cost position and persuade know-how dependent companies to stay or make new investments.

SUMMARY

Success is virtually independent of location. Even a manufacturing-cost disadvantage is not an unalterable feature of location. All the same, the laggards have a particularly large gap to close here, and will have a particularly long and arduous way to go to catch up. Unit-cost differences of 30 to over 50 per cent are a clear enough indication of this.

- A good half of the manufacturing-cost disadvantage (about 25 percentage points) can be eliminated if product and process design are better tailored to manufacture, a further 10 to 20 percentage points can be derived from greater operational efficiency. Only 5 to 10 percentage points are factor cost differences and thus largely outside management control. These must be compensated for – perhaps even outweighed – in high-wage countries by more innovative products and more intelligent processes.
- Manufacturable design allows both higher quality and more stable processes at much lower cost. Systems support is helpful here, but the most important lever is more intensive cross-functional collaboration. Manufacturing should not be overburdened by too many different factors.

- Successful companies are never satisfied with one-off redesign and do not bind themselves by the heels with excess automation. Continuous small improvements help them to reach their target of 99 and at times 100 per cent error-free production. Currently less-successful companies aiming to catch up must think more about better exploiting the potential of their technical plant and their workforce, instead of trying to fulfil the requirements on the manufacturing function with systems, robots and automatic machinery.
- Top companies use their production 'architecture' to improve performance systematically – by concentration and capacity adaptation within the existing configuration, and also by the targeted choice of new locations offering favourable factor costs, access to attractive markets, or a route to attractive innovation potentials.

5 Successful Innovation is Measured in Yield

Focused product features that satisfy customer needs without expensive extras, simple designs and processes – all the cost factors in manufacturing that influence corporate performance (for good or ill) mainly lie outside its own sphere. The foundations of innovative solutions (or the lack of them) are laid up front by research and development. And the mark of successful companies here is their ability to achieve better results at considerably lower cost.

They have superior products and lower unit costs, and get to the market earlier with new developments. Successful electronics companies have better control than their followers over both efficiency – that is, development productivity – and effectiveness – that is, development yield.

Taking development spend as a measure of productivity, we found that successful companies invest nearly 30 per cent less per dollar of sales in product development. In all segments, these companies spend less than 10 per cent of sales on R&D. Despite this economy in both relative and absolute terms, their design engineers generate two to three times the sales per capita from new products than less successful competitors do (see Figure 5.1).

Successful companies also put significantly more new products onto the market per development engineer. In computers/communication the figure is 35 new products per 100 engineers a year, higher by a factor of 3.9; in industrial electronics/measuring systems the figure is 3.1 times as many with 22 new products.

How do they do it? Their extremely targeted and highly productive innovation process reveals common features which can be grouped into three success factors.

- *Project portfolio: yield as the ultimate yardstick.* The self-actualization of technology-driven design engineers and the marketing manager's dreams of total market coverage do not figure when successful electronics companies come to define their development

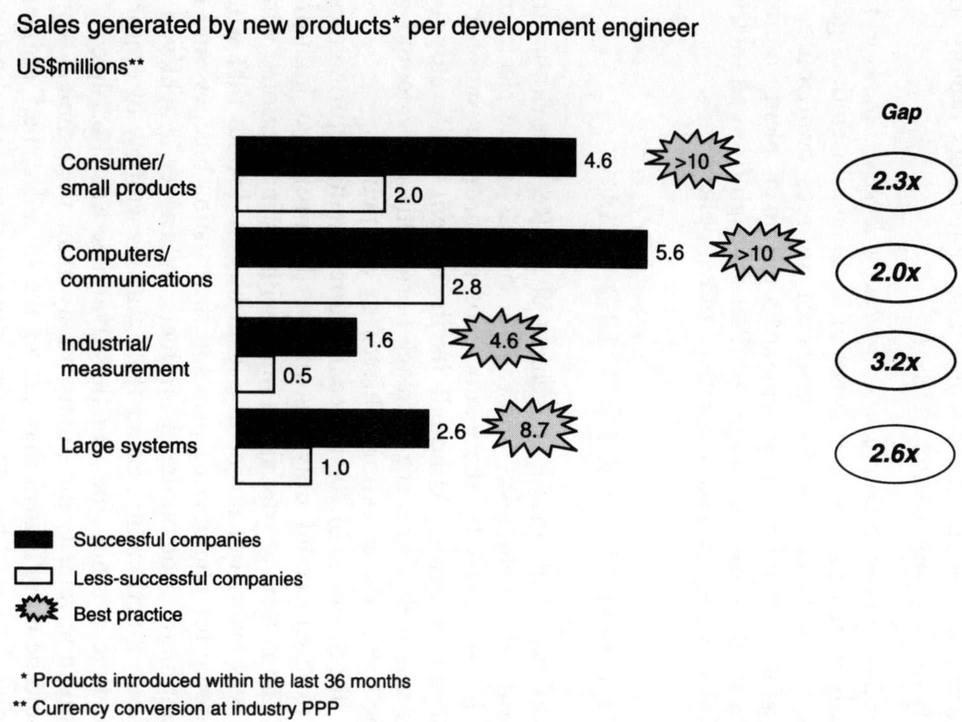

Fig. 5.1 *Development productivity two to three times higher*

projects. What counts is the yield in customer value and internal productivity.

- *Development process: integrated and flexible.* The risk of developing past the market is eliminated by a development process that ensures the maximum integration of all concerned – customers, suppliers and in-house functions – and that reacts rapidly and flexibly to changing market needs.
- *Innovation-friendly environment: stimulation density is the key.* The best way to generate an abundance of ideas and a wealth of good results is to have a wide variety of sources contribute. In successful companies, design engineers are given plenty of freedom, and ideas are encouraged to filter through from all internal functions as well as from customers and suppliers.

PROJECT PORTFOLIO ORIENTED TO YIELD

The importance of planned management of the R&D portfolio, that is, of the development tasks in hand, to success in the electronics industry is illustrated by the example of a European supplier of electronic components. Although its investment in development resources was well above the industry norm, this player could not match the speed of innovation achieved by competitors.

A cost disadvantage of 100 per cent in direct manufacturing meant that its products could only be sold in the protected high-price market. The company was also failing with its innovations, which meant it was losing out on income from new businesses that would have made up for plummeting prices in the existing business and for the discontinuation of outdated products. Whenever it achieved an advantage in functionality, either there was no window of opportunity – the time was not ripe – or the advantage was short-lived.

Steps taken to remedy the situation only made the problem worse. What could be more logical than taking the so-called 'bad' segments and 'sorting them out' by adding development projects and optimizing product cost? After that, a task force of design engineers and technologists would have to take on the extremely uncompetitive products in the loss-making segments and redesign them completely. The investment would surely pay off very quickly.

Unfortunately, however, things do not work out as easily as that, as similar cases have often shown. Viewed in isolation, such a programme would undoubtedly deliver the promised goods. But where are the required resources to come from? Either the research and development budget would have to be increased, making the overall cost position worse, or resources would have to be diverted from other segments of the business with high core competence and good earning power. But the industry dynamics will not tolerate this. The few 'good' segments need as many development resources as possible to keep pace with competitors' speed of innovation and maintain their respectable cost position by redesign and design-to-cost projects.

The 'repair strategy' – draining resources from successful segments – is almost certainly doomed to failure. The subsidized segments will make progress; but they will still trail behind already successful competitors, who will continue to improve at undiminished speed. At the same time, the 'good' segments will lose ground, unable to sustain the loss of resources.

It is impossible to keep too many balls in the air at once. Maximum yield from a development portfolio is gained in other ways. The most important of these are: focus on a small number of projects, restrict in-house input, and define business and project goals realistically and clearly.

Focus on a Small Number of Projects

Accelerating innovation in one product segment eats up additional resources. A company can only reach high speeds by cutting out all projects that lag behind the competition, and putting the lion's share of its resources into the 'good' segments. Experience shows that the inevitable risks entailed in such focusing are much less serious than the certain losses caused by fragmenting resources.

This can be illustrated by the example of a small manufacturer in the microcontroller business. The company was trying to emulate industry leader Motorola, and served the same number of application segments. But for a satisfactory cost structure in this business, development costs should not exceed 10 per cent of sales. This company had only one-fifteenth of Motorola's sales. So if Motorola used its full 'allowance', it could put 15 times more resources into developing each of its architectures and application segments – and its lead could grow ever wider. In the long term, the small competitor's only chance of survival is to cut back radically its number of architectures and concentrate on application segments where it has a particular core competence.

The long-term study confirms this success formula of concentrating your forces on a small number of projects. Successful companies have a much less complex portfolio of development projects. And since they manage with significantly fewer variants in their range of products, the number of sub-modules and parts requiring development is also reduced.

On average, successful companies therefore have about half as many projects running at the same time (Figure 5.2). Since their development resources are not fragmented, they achieve both greater funding per project and higher levels of project success with lower overall development expenditure.

Restricting In-House Input

Wherever maximum speed of innovation is a must (and that includes the electronics industry), very few companies can afford to run their entire development programme themselves. The successful ones have taken the lead here, too. They long since started confining their in-house development input to key tasks, and outsourcing those parts of development that do not provide competitive differentiation and that are relatively simple to define.

To get fast enough and keep getting faster, today's successful companies have rephrased the make/buy question. It is no longer 'Which operations are unimportant enough to be outsourced?', but 'How can we achieve the greatest advance in innovation in the shortest time?' If, using its own development resources, a company will arrive too late to exploit a window of opportunity, which is not uncommon, then a solution has to be bought in from outside if possible. In such cases, successful companies put aside traditional doubts about dependence and loss of know-how. Hewlett-Packard outsourced the printer head of its globally successful printers right from the start.

In order to achieve the desired efficiency gain and concentrate in-house energies, successful companies give their subcontractors detailed instructions as to the necessary features and deadlines. They promote efficient teamwork by bundling individual parts into sub-assemblies or modules that can be bought in as a single unit. One manufacturer gave up in-house development and manufacture of

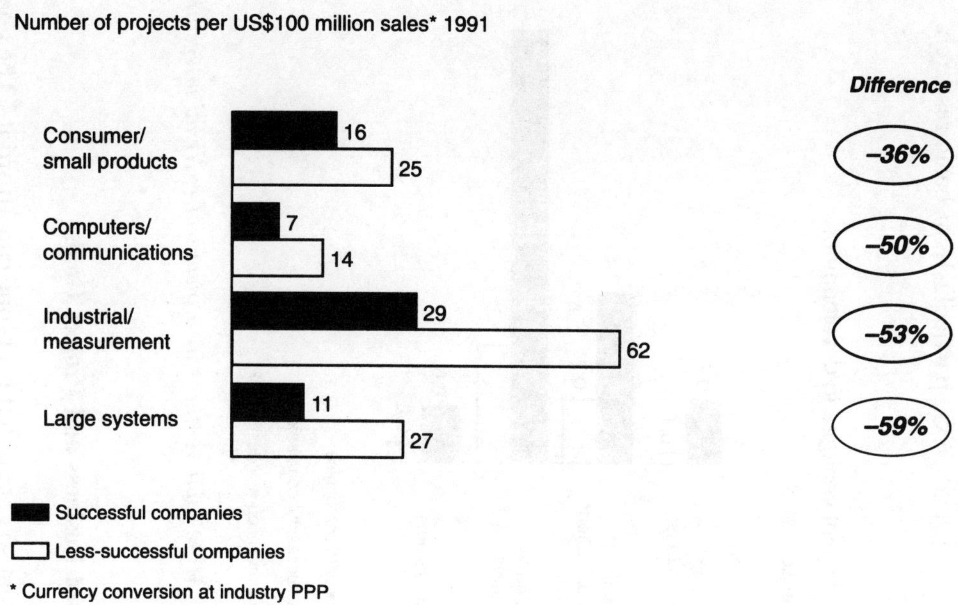

Fig. 5.2 *Fewer development projects in parallel*

entire SMT subassemblies, reducing the number of components that had to be managed in-house by over 90 per cent.

Another form of work-sharing that successful companies use to good effect is development cooperations, which they use on average three times as much as less-successful companies (see Figure 5.3). These joint ventures allow them to tap into critical competitor know-how and help to spread risks, particularly in basic research.

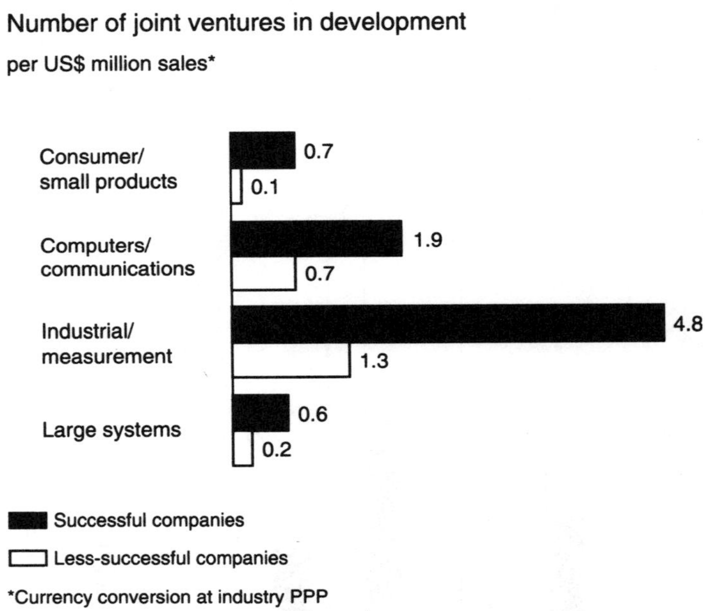

Fig. 5.3 *More joint ventures as a means of external integration*

Clearly Defined Business and Project Goals

'A mass product so cheap that everyone can afford it.' The goal set by Nintendo's boss, Hiroshi Yamaouchi, for his development engineers and game designers at the start of the Famicom project could not have been more clearly put. More importantly, it simultaneously defines the nature of the business and the outcome of the concrete development project in hand. These are, again, two essential factors, particularly at extremely high rates of innovation.

Without a precise, shared understanding of the business, the much-invoked entrepreneurial thinking and action – and with it effective innovation strategy – remains an illusion. Even skunk works, often praised as a panacea for improving innovativeness and shortening decision paths, are not much help unless all participants can agree whether they are in a systems integration business, a hardware or software business, or a combination of the two.

Successful companies gear their innovation strategy to the success factors of their businesses. The industry segments surveyed in the long-term study can roughly be split into two here. The consumer-electronics and computers/communications segments win in the market on price, while industrial electronics and large systems depend on product performance and customer value over the entire product life cycle.

In the commodity business (for example, standard PCs, VCRs), where successful differentiation is practically all price-based, innovation is mainly targeted at low-cost production. (Of course, it is not impossible for players in these segments to differentiate for a short period by superior customer value, as was recently demonstrated in the case of direct speech input for car phones.)

With highly specialized systems, such as industrial control technology, improved product performance has higher priority. There is often much more to be gained on the user side by electronics-aided improvements in yield or productivity than by mere reductions in the amount of investment. If a process's spoilage rate can be halved by using better control systems, and if the process yield is increased thereby from 90 to 95 per cent, the price of the new control technology – if it is compatible with the existing plant – will be of secondary importance in the buying decision.

Successful companies set their priorities accordingly. In the consumer electronics/small products and computer segments, their high sales are due to cheaply-priced products, not to technological superiority. In all other segments, the players owe their success to a high share of sales of technologically-superior products.

The aims of development projects are in line with this understanding of the business. Among successful companies, particularly in the consumer electronics/small products and computers/communication segments, the results can be seen in an innovative product range. Nearly half the products are launches from the last twelve months, compared with only a quarter for less-successful companies (Figure 5.4).

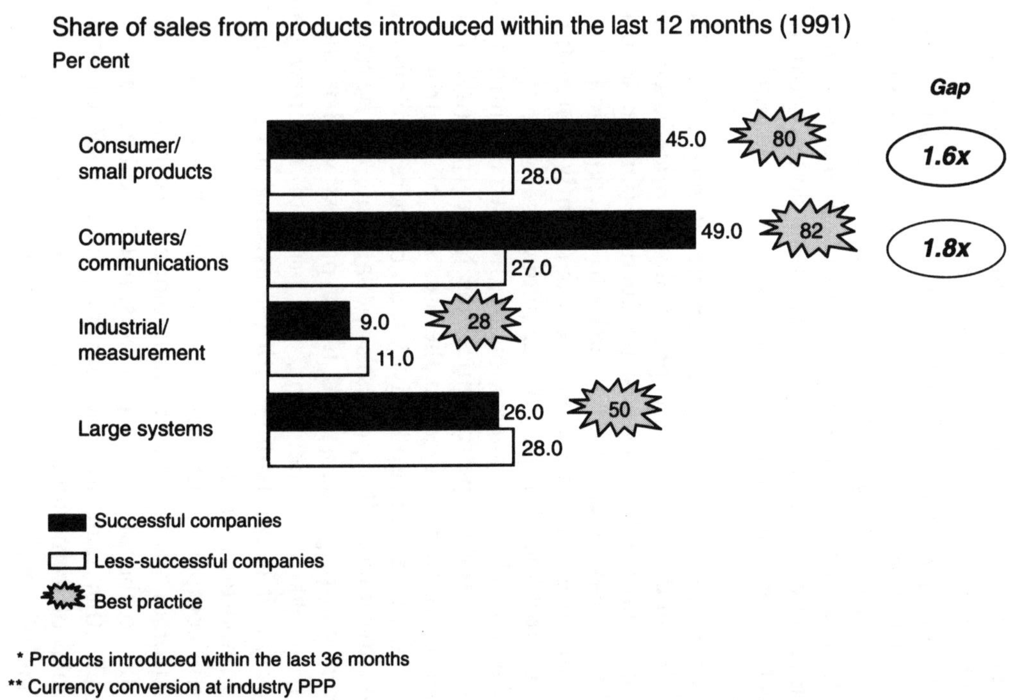

Fig. 5.4 *More sales from first-year products in consumer and computer segments*

Enormous savings can be achieved with an innovative product range. A rapid succession of product redesigns from development regularly gives companies the opportunity to exploit the latest progress in component integration to reduce costs. Using these tactics, the most successful companies surveyed in the long-term study were able in 1991 to cut their product costs by 40 per cent more than less-successful players.

By contrast, in the technologically higher-value segments of industrial electronics/measuring systems and large systems, highly-innovative product ranges are not essential to success. Successful companies in these segments have clear product advantages that will endure for some time. All the same, successful producers of industrial electronics and measuring systems are first on the market with their innovations more than twice as often as less-successful companies. As innovation leaders in a business where product performance is the decisive buying factor, they can put high volumes on the market at high prices and skim substantial profits.

INTEGRATIVE AND FLEXIBLE DEVELOPMENT PROCESS

It is widely accepted that product development cannot be left to the development department alone. But the gulf between theory and practice can still be great, as shown by the – by no means unique – example of one measuring-systems manufacturer.

On the basis of previously unapplied physical principles, the company developed measuring processes of unprecedented precision at great expense. It turned out that the total value of the market for these new products came to almost exactly the same amount as the development spend. A technologically-successful development became a commercial flop. Having neglected to find out how big the relevant market segment was before starting development, the company had developed its product right past the market.

Another development led to major problems at production start-up. After developing one product past the market, the company was very careful to match the product definition to customer requirements. At great expense the company developed a concept for a new product that was intended to give the customer significant additional value. This was to be achieved particularly through high stability in the calibration of the instrument, which would eliminate cost-intensive servicing.

Having successfully manufactured a limited selection of pilot products that covered only a small measuring range, the company then heavily promoted the new product range covering all measuring ranges, promising early delivery. The concept made sense to customers, and orders poured in. Unfortunately, however, the company encountered severe problems in manufacturing: on some production lines only 30 per cent of units were perfect on first pass, and 70 per cent had to be either scrapped or repaired at high cost, while in some stages of the manufacturing process entire batches were often faulty.

The problems in the production process proved complex and stubborn. Even a year later they had not been completely resolved. Meanwhile, of course, substantial delivery problems arose, causing customers and, worse still, the company's own salesforce to lose confidence in the new product that had been developed at such great expense. These major start-up problems jeopardized the commercial success of the entire project.

The product had been developed in the development department, without sufficient involvement of manufacturing experts. Once the development work had been completed, the product had been thrown 'over the fence' to manufacturing – where development had had to start all over again. Problems in the manufacturing process could have been identified and solved much earlier, during development, if only manufacturing had been sufficiently involved at the time.

Development risk stems from misinterpretation not only of market conditions, but also of in-house requirements. The only solution is genuine involvement of all internal functions concerned and systematic integration of customer demands.

Extremely Close Functional Integration

Successful companies make provisions for the early involvement of all relevant functions. In the strongly cost-oriented computer segment these are manufacturing, purchasing and accounting. In the customer value-oriented segment of industrial electronics/measuring systems it is usual to involve customer service at the concept stage in the interest of manufacturing products that are easy to maintain and service.

The organizational solution to this cross-functional cooperation is increasingly found in specialized 'product pipelines' targeted to market segments and including all functions needed for the success of a product. Market, development, manufacturing and service needs can be better considered in these small vertical units with short feedback loops than in a function-based matrix organization.

In these units, product management adopts the role of the 'entrepreneur', no longer serving as a staff function of the marketing department. Suppliers who develop and produce individual product elements independently are also fully integrated into these vertical units.

The argument still advanced against these vertical units is the loss of synergy. But you can only lose what is actually there. The fact is that the synergy potential – in the sense of customer loyalty based on the offer of 'one-stop shopping' – generally exists only in the supplier's imagination, and the customer's interests are quite different. The additional one-stop-shop value generated when the customer can choose from a broad product range at a single supplier typically carries a 5 to 10 per cent price premium, normally well below the costs associated with the additional complexity.

To the customer, a focused range geared to his needs is the most important thing in an 'open environment' where components are becoming increasingly compatible. The customer likes having the opportunity to build a system with the best components from various producers. The one-stop shop is something fewer and fewer customers want. Not many are willing, especially with large systems, to place themselves at the mercy of a single producer.

So companies with product pipelines are doing exactly the right thing. They are abandoning out-of-date synergy conceptions and gaining respect and trust from their customers because their focus gives them superior customer-specific competence.

Systematic Integration of Customer Requirements

To be able to feed customer preferences into their design specifications, successful companies use such tools as 'Quality Function Deployment' (QFD). This process first assigns customer requirements to the main features of a new product, and then translates these into technical requirements.

Kiyoshi Uchimaru, president of NEC's microchip subsidiary NIMS, solved his problem with chip design this way. His engineers had got some specifications wrong in the development of the NIMS 16-bit chip – not many, but ones that were important to the customers. As a result, the finished chip did not have the desired success in the market. After this experience, Uchimaru decided to put his faith in the QFD system. When developing the NIMS 32-bit microprocessor, he and his team drew up matrices which they could use to determine the correlation between customer demands, technical specifications, hardware design characteristics, and every function required. This process helped identify potential design problems at an early stage, locate design bottlenecks, and define the limits of the design. The result speaks for itself: every specification was met, and there was a sharp increase in the productivity of the development process at the same time.

With this approach, development engineers can target customer requirements while avoiding expensive subsequent modifications. In consumer electronics, QFD is used twice as often by successful companies as by less-successful ones. In the industrial electronics/ measuring systems segment, about half of all successful companies use this method, while the less successful have yet to discover it.

Systematic identification of customer preferences is followed by product definition. Here, successful companies go further than ensuring that the product meets the needs of the envisaged customer; they also keep a check to ensure that the search for customer value is not turning into a costly fragmentation of effort and excessive variety. In each segment, the advantages of individual product features must be carefully traded off against their additional costs. In the computers/communications and consumer electronics segments, where minimum cost is the deciding factor, the lowest possible variety is desirable.

Superior Project Management

In the course of individual development projects, duration is not really the factor distinguishing good companies from bad. Project time is practically the same for comparable products as shown in Figure 5.5; in some cases the less successful players are actually faster. Where the major differences are to be found is mainly in project management.

Successful companies allow themselves only a very short interval between finalizing specifications and launching products on the market, so they have as much time as possible to respond to market requirements. The 'design freeze' for PCs, for example, is about seven weeks before market launch in successful companies and more than five months at less successful ones (Figure 5.6).

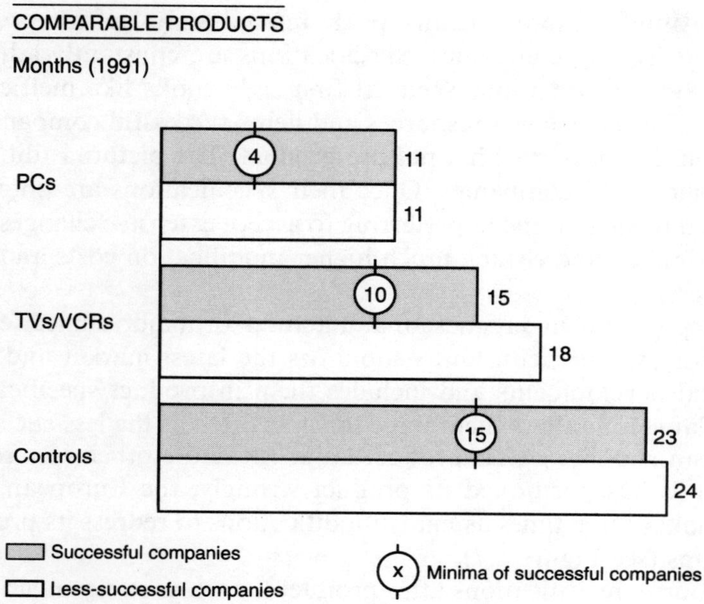

Fig. 5.5 *Development times hardly differ*

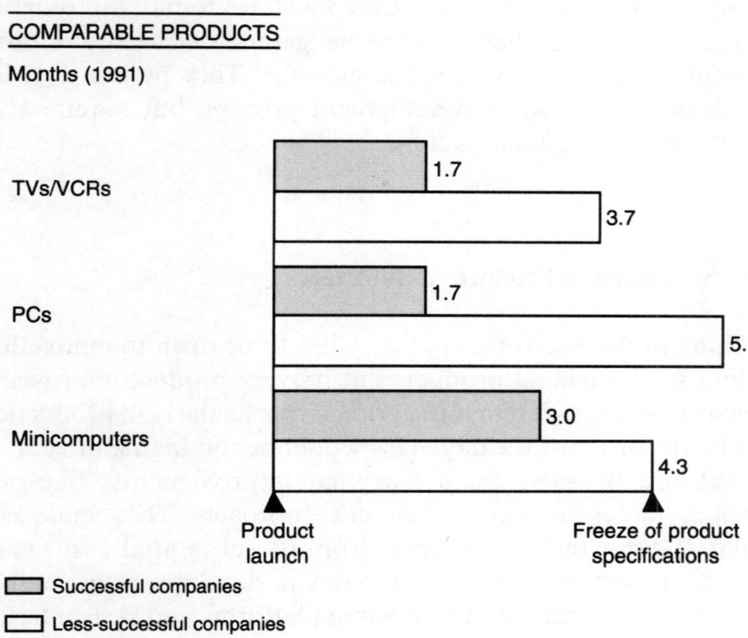

Fig. 5.6 *Short time from design freeze to product launch*

Accordingly, modifications peak immediately before product launch at leading companies. Specifications are constantly adjusted to the latest information. What at first sight looks like inefficiency actually consumes fewer resources and helps successful companies to arrive on the market with a mature product. The picture is different at less-successful companies. Once their specifications are approved they tend to stick to them, preferring to accept extensive changes after market launch and risking much higher modification costs and irate customers.

A very successful Japanese manufacturer of major clinical equipment, for example, constantly monitors the latest market and technological developments and includes them in product specifications before launch. It alters them three times as often as the less-successful European supplier. After product launch it is the other way round. Because it has positioned its product wrongly, the European company makes three times as many modifications to redress its previous omissions (see Figure 5.7).

Of course, modifications after product launch are very much more expensive. Tools have to be altered, the production process adjusted, service staff retrained, and new documentation drawn up. New suppliers must be sought, new uses must be found for otherwise unusable stocks: the changes run the gamut of the organization. Successful companies have got the message. They permit high flexibility during the product development process, but require stable design when starting manufacture.

Robust Modularized Product Architecture

Everything in the electronics industry has to be open to innovation – including the individual product. But if every product improvement involved complete redesign, disruptions – particularly in production – would be the order of the day. This would be too high a price to pay for what are in most cases marginal improvements. Successful companies therefore take a different approach. They build on a modular concept that, quite apart from its substantial cost benefits to manufacturing, helps boost efficiency in development as well.

Modular structures with unchanging platforms and clear interfaces across several product families are a particularly good way of

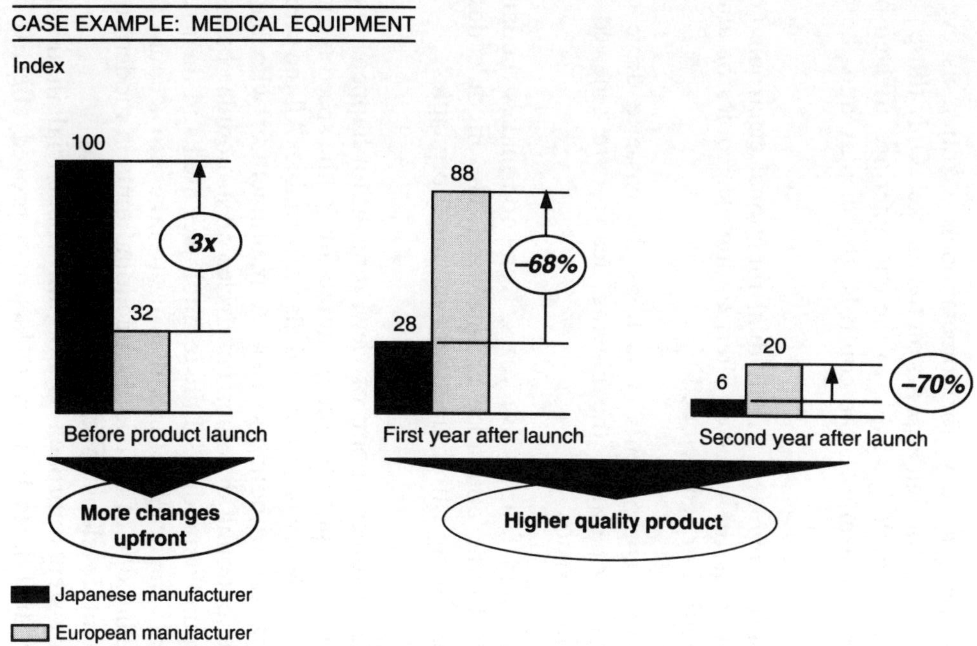

Fig. 5.7 *More specification changes upfront reduce subsequent changes by two-thirds*

integrating new technologies and extending functionality. Individual modules are updated step by step and the product life cycle is paralleled by module and technology life cycles.

The Kenwood car radio cassette recorder provides an example of four such life cycles:

1. At least once a year, for some models as often as every six months, a new design appears on the market. Only those parts of the recorder directly visible to the customer are affected by this redesign, for example, the controls, the display and the casing design.
2. About every two years individual functional electronic components such as the receiver filter, tuning electronics or amplifier are modified or redeveloped.
3. Mechanical components – such as the cassette drive or the mechanism operating the playing head – are renewed about every three years.
4. Finally, every four to five years, Kenwood makes fundamental changes to the appliance architecture. The module design is reviewed, along with individual functional principles.

From the customer's point of view, innovation happens fast with this modular concept; from the manufacturer's, the speed is manageable since complexity is limited and also affordable. All the same, this procedure is not without its risks. A manufacturer who bases his design on uncoupling development cycles on a modular basis needs an in-depth understanding of the technologies used and their remaining potential. The design of modules has to take into account the fact that the technology of individual functional units in modern electronic equipment develops at different speeds.

A step-by-step approach is not only beneficial in developing existing product lines further. It has also proved effective in the transition to completely new basic technologies. An example here is Northern Telecom. Instead of changing the whole range of telephone equipment from analogue to digital at a single stroke, the company started by using digital technology in the simpler PBX systems – and then only in a sub-function: only after a successful 'trial run' were the larger switches progressively digitalized.

This 'bite-sized chunks' approach allowed Northern Telecom to bundle its resources – the individual product was faster to market. But not only that: the experience gained at each stage was fed back into the development work. Development cost could be minimized without compromising on speed. The result was that Northern Telecom had its fully digital PBX generation much sooner and much cheaper than any of its competitors.

The combination of step-by-step innovation and the modular principle allows concentrated use of development resources. Development engineers are no longer overloaded with projects, and are therefore no longer a bottleneck. As the number of projects falls, the productivity of development engineers rises. And the development portfolio is easier to manage too: costly interface adjustments are no longer required and the trail of changes lying in the wake of every modification becomes easier to control.

INNOVATION-FRIENDLY ENVIRONMENT

A strictly yield-oriented project portfolio and an integrated, focused development process go a long way towards turning the development function into a wellspring of corporate productivity and growth. The flow from this wellspring – whether it will gush out a constant stream of improvement ideas so that a company drives the innovation race instead of being driven – is decided elsewhere: in the type of organizational setup used. The organizational framework for R&D may either ensure maximum internal and external stimulation and a highly fruitful climate for innovation, or, at the other extreme, lead to isolation and atrophy.

The best companies have developed the innovation-friendly environment to perfection, mainly by diligently observing three requirements:

- Development teams must be given sufficient independent responsibility to secure maximum flexibility and responsiveness;
- They must be staffed with the best people to ensure that the best use is made of the resulting freedom;
- And the choice of location must be deliberately targeted at strengthening the skills-base and stimulation density for innovation.

Independent Development Teams

Any company would categorically deny that it failed to exploit the advantages of project teams in new product development. But such claims are not always borne out by reality.

True, everyone knows that project organization is particularly well-suited to development projects, whereas line organization tends to promote conflicts of interest and often wrong behaviour. But, in the past, so-called project organization has often amounted to little more than conventional working groups. Development engineers still find themselves faced with the dilemma of having to please both line management and development coordinators. Some development projects still go on and on – products continue to be improved long after market introduction because the resources assigned to them have not yet been reallocated to new projects. This not only eats up development capacity and unfocuses the development portfolio, it also burdens the whole company with huge modification costs.

Project teams that have to work this way bear more resemblance to loose working parties, yet they are more widespread than is generally assumed. The long-term study found, for example, that in the consumer electronics/small products segment only two-thirds of all successful companies met the essential requirements for efficient project organization:

- Full-time development teams with a fixed assignment; and
- Project managers with sufficient authority.

Among the less-successful consumer electronics manufacturers, not one has both these characteristics.

Dedicated development teams are formed for a special assignment and broken up again when it is over. Their members are permanently appointed and are available full time. The compensation systems for these teams are geared to the market success of the products and not to internal parameters such as on-time completion or tenure with the company. And these teams are, of course, not made up of development engineers only. They include employees from all functions on full-time secondment: marketing, purchasing, production, service, sales and – wherever possible – representatives of the customers.

These project teams are headed by managerially responsible project managers with extended authority. They have direct access to top management and a say in the choice of their team members. This is another area in which successful companies are further ahead than their less-successful competitors: more than twice as many of their project managers report directly to top management; they are twice as likely to have the last word on the project budget, and are involved in decisions on the staffing of their teams five times as often (see Figure 5.8).

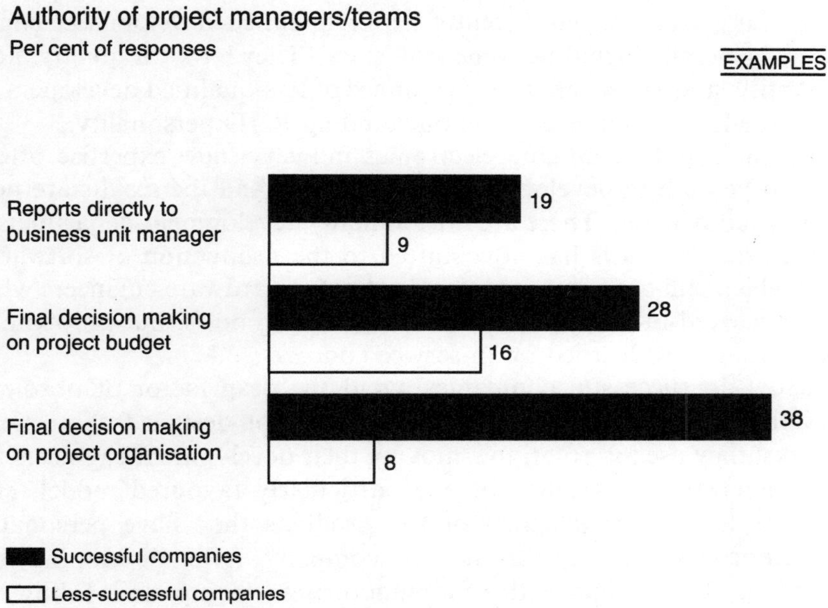

Fig. 5.8 *Project managers and teams in successful companies have more authority*

Granting teams such independent accountability is the only way to facilitate flexible and fast response to market demands. Changes in direction can be made without long coordination loops. With customers, sales and marketing involved, important market trends percolate straight through to the team, and the different functional points of view make for a variety of creative inputs that could never be encountered in a pure 'developers' club'.

The less-successful companies will learn from their successful competitors here too. They will gradually reject their loose working parties and turn to small, cohesive teams. And it is to be hoped that they will also remember another lesson from the successful companies: the best team structure will do no good if the staffing is wrong. The main success factor is the selection of the right people.

Top People on the Team

On the whole, the development departments of successful companies have far fewer, but consistently better qualified development engineers, generally including some real 'stars'. They look for quality, not quantity, and know that a large number of less-qualified developers is not worth as much as a single outstanding R&D personality.

In the rapidly changing electronics industry, new expertise often has to be built in development departments. And the results are not always convincing. There are far too many development departments where the emphasis has now shifted to the production of software, but which still contain a large proportion of hardware engineers who are not dyed-in-the-wool software engineering buffs, but only know what they have learned on in-service courses.

How do successful companies avoid the drag factor of obsolete skilling? They offer career paths that are not designed as vertical stacks; they use job rotation – moving their development engineers to manufacturing and sales. In one particularly favoured model, engineers follow the progress of the products they have personally developed all the way through the company. After developing the product, they prepare the manufacturing startup and follow it through to series production. Only the very best people are then allowed to return to development. An important side effect is that the development team stays young.

Of course, top developers can only be attracted and retained if there is a good infrastructure in place. And that is why successful companies spend on average 30 to 90 per cent per development engineer more on infrastructure, depending on the segment (see Figure 5.9).

Building such personnel resources and attracting the development stars is a long process. What can be done if the 'window of opportunity' for a particular development is going to open in two years, and the company cannot build the necessary know-how in time?

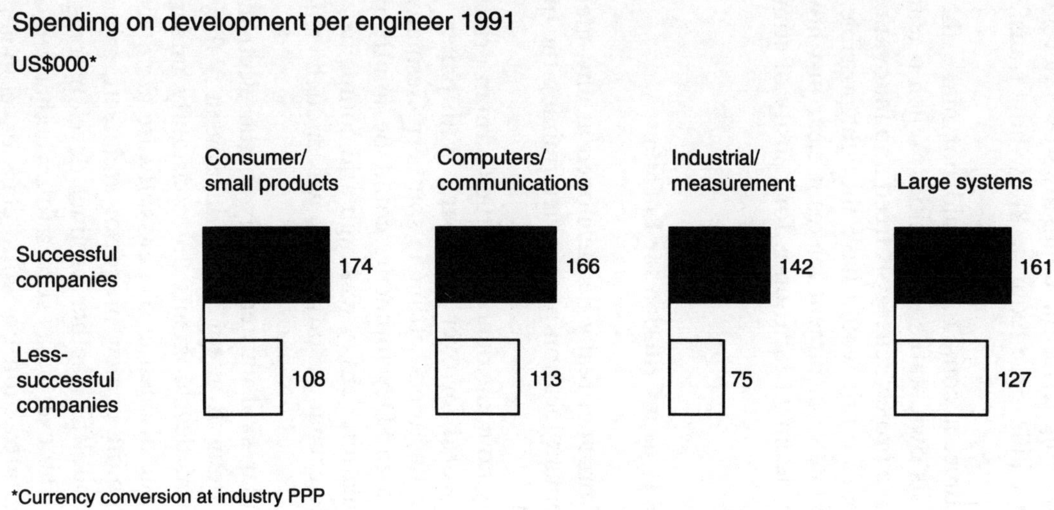

Fig. 5.9 *Better development infrastructure*

In this situation the successful, innovative company will go where the resources are. For example, it will move its development department to Silicon Valley or import people with critical know-how from the US. It will then have a better chance of business success in the short term. In the long term, the company's own development team must acquire new skills – the requisite know-how transfer is the first step here.

At the same time, a company should not make the mistake of underestimating its own staff's capabilities. When a team can focus its attention on one project, energy is freed and innovation speeds are often attained that no one would have thought possible.

However, this line of argument makes it clear just how important the choice of development location is for density of stimulation.

Stimulation Density as a Criterion for Location

Can all types of product really be developed at any site? What part does the development location play in determining the quality of the new development?

It was certainly no coincidence that the German printing-machine industry led the world for over 100 years. With paper, printing ink and machine manufacturers settling in southern Germany, skills were bundled in one area. Developments could be jointly planned and tested. The stimulation density was high and companies could easily communicate and exchange staff. There was enough critical mass for innovation.

There are similar skill clusters today in the electronics industry around Boston along Route 128 and in Silicon Valley, where the latest technological developments are constantly under discussion, even at parties. The discussion of new software architecture, object-oriented programming or new processor architectures and models is universal. It is doubtful whether product developments away from the main centres can ever be as successful, whether remote locations exercise the same attraction on world-class development engineers, and whether the critical mass of specialist discussion partners can ever be attained elsewhere.

Alongside high skill levels, an innovation-friendly environment features innovative customers and competitors. Conservative customers do not generate a favourable climate for innovation and cannot give a reliable indication of market trends. Whether a company operates in a closed market defined by proprietary standards or in an open, competitive environment with aggressive competitors also makes a huge difference. The one may seriously retard innovative energy, with no incentive for renewal; the other demands continuous innovation for survival.

In many cases, innovation is stifled by statute. The extremely tight tolerances that must be observed in Germany when calibrating certain flowmeters can currently only be met with great difficulty if electronic–physical measuring principles are used – so the German market's standard model still works on the mechanical principle developed around the turn of the century.

Other countries do not require such extreme accuracy, with the result that electronic–physical principles have prevailed for some time because of the lower maintenance requirements and the very much lower purchase price. Manufacturers concentrating on the German market have been unable to innovate. Meanwhile, the new measuring methods are on the threshold of the accuracy required for Germany, and manufacturers who have clung to the mechanical measuring principle for too long run the risk of losing their core market to very much cheaper competitors.

Innovative customers ensure manufacturers a high level of stimulation density. Presence in innovative lead markets guards against the risk of missing important innovations. It is particularly important for European companies in the electronics industry, with their somewhat conservative customers, to expose themselves to a more exacting and therefore more favourable climate of innovation. That is why expansion into innovative lead markets such as the USA or the Far East must be pushed as a top priority. And the learning experience in these markets must be used not only to maintain innovation clusters in Europe, but also to build more.

SUMMARY

Their development departments give successful companies a superior yield of newly developed products, with features closely modelled on current demand at the point of product launch. They achieve this – as might be expected given the high pressure of costs – not by a particularly high input of resources, but by outstanding efficiency in their development work, for which they rigorously establish the right conditions:

- In their *project portfolios* they concentrate on a small number of high-potential projects: they limit their own input to the most important development stages, and outsource even these if market success depends on it or if their own resources can be used more productively elsewhere. And, depending on the character of their business, they set unequivocal and ambitious cost or value goals for each project.
- In the *process of product development* their particular strength does not lie in shorter project times – where they have little advantage over the less successful – but in achieving maximum integration and flexibility. Both customer demands and the experience of the other functions in the company are systematically involved. And right up to market launch they demonstrate extreme flexibility and tolerance to change. From the point of series production onwards, however, their freedom from defects and stable design leaves their less-successful competitors far behind.
- Finally, by creating a particularly *innovation-friendly environment* the successful companies provide their development engineers with maximum stimulation density in every conceivable way. Their project organization consists of independently responsible development teams containing top personnel and led by a managerially-responsible project manager. And their development locations are selected for their stimulation density – the skills base, demanding customers, aggressive competitors, and legal regulations that foster innovation are the main criteria.

6 New Businesses Must Hit Windows of Opportunity Just Right

Volume accrues even faster than plummeting prices erode value-added. Unfortunately not in every company, but certainly in the electronics industry as a whole. The tumbling prices contribute to the surges in volume, since they create the mass markets. But the most powerful drivers of growth and revenue are new businesses. Companies planning to profit from these need more than products with innovative customer value. They have to allow for and shape conditions beyond their own sphere of activity.

Billion-dollar businesses dominate the scene in the electronics industry. A few years ago, most of them did not even exist. From personal computers, camcorders and compact disc players to digital switches and electronic motor management systems or computer games – over and over again, the electronics industry creates significant new customer value from new technical capabilities.

The European electronics industry had no founder members in most of the new businesses. Europe did not pioneer segments such as personal computers and computer games, and elsewhere, as with video recorders, it came first too soon (Philips video recorders years before VHS and Betamax) and then too late (Philips/Grundig Video 2000 after VHS and Betamax). And so European companies have lost substantial revenue potential that could have financed the succeeding technical generations.

The reasons are not due to insufficient investment in technology – in relation to sales, European manufacturers have invested at least as much as their American and Japanese competitors. Nor is it true to say that European companies were not quick enough to foresee the new applications and businesses of the future. Opportunities were most likely lost because the talents of universities and industry were not properly bundled and deregulation, for instance in telecommunications, came too late.

It also seems that European companies, after long years of success in traditional business sectors, could not cope with the requirements of the youthful electronics industry.

The enormous dynamics of innovation are far more challenging here than in other industries: new generations of technology in microelectronics every three years or so, a two-fold increase in the technical performance of processors about every two years, and drastically shortened life cycles in consumer electronics and PCs. This factor is now widely understood and, in recent years in particular, has led to huge efforts, some successful, to catch up with the leaders.

But a far more decisive factor in this industry appears to be the fact that new products have to target a very small window of opportunity. Many new applications entail substantial changes to the existing environment. Infrastructure investments are often necessary for a new electronic product to be able to develop its full value; new networks need to be installed, software changed, and new interface standards agreed. And often a certain critical mass is needed before a product's full value can unfold.

This makes electronics an example of an industry with a positive feedback structure, as described by American economist Brian Arthur (1989):* a highly non-linear entity where, in contrast to traditional industries, the sum of the forces influencing business success (for example, customer value, cost structures) does not lead to a clear, predictable state of equilibrium. Whether a company will succeed or fail may depend on the tiniest changes in the business environment.

How the combination of these two factors – extreme innovation dynamics combined with heavy dependence on positive feedback from the environment – can produce hits with undreamed-of growth rates is impressively evident from some recent success stories. But even top companies like Nintendo or Sony have demonstrated equally memorably how neglect of the business environment can ruin the apparently most promising startups. Clear lessons for today's laggards can be drawn from both the hits and the flops.

---

* Arthur, W. B., 'Competing technologies, increasing returns, and lock-in by historical events', *Economic Journal*, March 1989, pp. 116–31.

NINTENDO

Measured by its profit and growth rates and by its market penetration at home and abroad, Nintendo is now one of Japan's most successful companies.* In the past ten years, this insignificant producer of playing cards and toys has grown into a multi-billion dollar electronics company with profits reaching $1.0 million per employee in 1993, albeit with extremely low vertical integration.

Behind the success of electronic child's play lay a carefully planned strategy. Nintendo played the success factors of the electronics mass market like a virtuoso musician teasing the best out of a concerto, using as its instrument a low-priced, high-performance gadget that seized a gap in the market at the right time and constantly stimulated follow-up sales with new games. Nintendo had – and has – no intentions of becoming a leader in the latest technologies; they do not matter to its target mass market.

Product development – which led first to the 'Famicom', which was not marketed in Europe, and later to the familiar 'Gameboy' – was to form the basis of Nintendo's vigorous expansion. Managing director Hiroshi Yamauchi set the ball rolling himself. He formulated very clear goals, based on customer value and possible competitive advantages, but not on technical criteria.

A product was to be created that would be safe from imitation by competitors for at least a year and that would be so good 'that even afterwards nobody will want to have a copy or an imitation, nothing but the Nintendo original will do'. The product had to be universally affordable. In 1983, the year planned for market launch, video and computer game consoles cost US$200 to US$300. Yamauchi set a price ceiling of less than US$75 for his 'Famicom'.

At the same time, the product's performance was to be much more powerful than that of its competitors. The rival products produced images and rows of digits, and could even perform complicated calculations – because they were developed by engineers, generally from the office computers sector. What many of them lacked was 'convincing' movement by the figures on the screen, as achieved by 'fast speed animation'.

---

* Sheff, D., *Nintendo Gameboy*, Munich, Goldmann, 1993.

Yamauchi, in contrast, hired engineers for his project who had experience of large amusement-arcade machines and who were used to thinking in terms of games. They designed a product that only needed an 8-bit processor – essential to meet the low price ceiling.

Yamauchi refused to compromise in the detailed design of the 'Famicom'. The game had to be fast so as to grab players' attention, and not bore them by slow computer response times. Special circuits had to be developed for this purpose. The 'Famicom' did not have a conventional keyboard, a modem or a floppy disk drive, and therefore would not be compared to a computer. Furthermore, because it did not use disks the games would not be easy to copy.

The search for sufficiently powerful chips for more realistic imaging proved a problem, though. Chip suppliers described Nintendo's ideas on price as completely off beam. They were finally induced to cooperate by a guarantee of a high order volume of three million chips in two years. At that time, not even Nintendo's best product, the TV Colour Game, had sold more than a million, and typical sales were more like 20 000 to 30 000 units a year.

Yamauchi then told major wholesalers that volume sales could only be guaranteed by very low prices, and that they would therefore have to take a lower mark-up than usual in this trade. He also explained that the more hardware sold, the greater would be the demand for software, which would be the real money-spinner. The wholesalers complied.

One absolutely critical success factor was the simultaneous provision of hardware and software. Separately from the hardware development engineers, design groups thought up imaginative video games and competed with one another for the best product. A major part of Nintendo's success was undoubtedly due to Miyamoto's invention of the Mario figure, which was popular even in American amusement arcades. The Mario games for the 'Famicom' were the most interesting ever to hit the market.

To make the gadget a long-term success, plenty of software would have to be available. Nintendo itself opted to excel with a small number of good games. Between 1985 and 1991, only eight Mario games came onto the market, though altogether 60 to 70 million copies were sold. Nintendo therefore operated a systematic licensing policy from the start. Licensees developed the software; but Nintendo handled production – remaining true to the principle that 'All "Famicom" games will be produced exclusively by Nintendo.'

With a minimum release of 10 000 copies per game, the licensees were geared to a high-volume business. And Nintendo was able to monitor game quality. In 1987 there were 17 licensees, a year later 30 – by 1991 their ranks had swelled to 90, and competition between game developers ensured that interesting games kept arriving on the market.

Despite its meteoric rise, Nintendo has remained a 'lean' and flexible company. Today it has 200 employees in research and development, 350 in administration, 180 in the main factory and 130 in the Oji factory – with 1993 sales totalling US$5.5 billion.

SONY, JVC, PHILIPS

The Walkman, camcorder and CD player have made Sony synonymous with new business hits. But the company also gained a certain notoriety from its abortive attempt to establish its Betamax standard in the video recorder business in the early 1980s. Despite its somewhat later arrival on the market, JVC's VHS system proved too strong for Betamax.

Technical performance differences were definitely not the critical factor. If anything, it was Betamax that had functional advantages. However, it soon transpired that the decisive feature in this business was not the 'video recorder' hardware, but the ensuing value that the customer could gain as a result of his purchase. The customer wanted to be able to rent videos and watch them at home, or play friends' recordings. Therefore, he would inevitably go for the recorder for which most videos were available, and not for a machine with a putative product advantage that the layman could barely grasp anyway.

JVC realized this quite clearly, and saw its chance in generously issuing licences to hardware suppliers. Sony, not wanting to give away its technical advantage to an unlimited number of competitors, was sparing with its licences and lost out.

The video rental companies, for their part, had no desire to pay the extra costs two systems would entail. They plumped for the standard system which sooner or later could expect the biggest market share, and that was VHS. By the time Philips and Grundig entered the market in 1982 with their technically far superior Video 2000 system, they were too late (see Figure 6.1). The switching costs both for rental and coding companies and for the end-user were huge, and could not be justified by a product advantage.

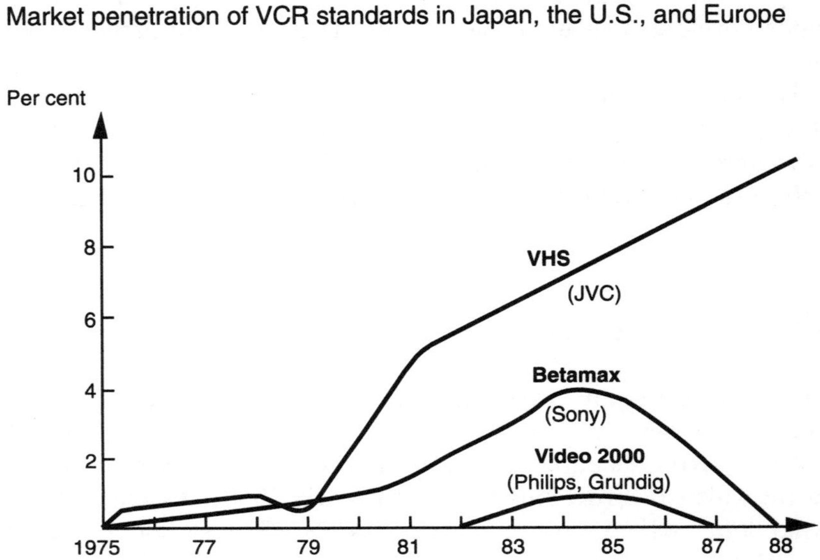

Fig. 6.1 *JVC's licensing strategy makes VHS a success*

Any advantage, however small, ultimately induces the whole body of customer behaviour to move towards a marginal alternative. In this case, the originally small lead in market share resulting from market feedback tipped the scales. Nearly 100 per cent market share for a system that started with only a slight edge was the result.

Both Philips and Sony encountered similar problems when they introduced the DCC and the Minidisc player respectively onto the market at the end of 1991. Philips's DCC digital cassette recorder is undoubtedly a major innovation compared with conventional analogue recorders. It achieves near CD quality with cassettes, and its shock-proof qualities make it easy to carry around. These features could make a product a huge hit, particularly since Philips had also given consideration to existing recordings made on the old analogue cassettes and ensured that they could be played on the new machine.

The minidisc player by Sony is similarly innovative. Its sophisticated optoelectronic technology means it can be used for recording, similarly to the DCC. Its interim memory facility makes it robust in mobile use. Accessing individual music tracks is easier than with the DCC. Another big hit in the making?

Both systems seem to be making little or no headway. This is not due to the numerous technical hitches that kept delaying market launch as much as to the simple fact that it did not suit software suppliers such as BMG, Warner and EMI to have another sound medium on the market – whether digital cassette player or minidisc – in 1992 when they were launched.

When CD players came onto the market in 1982–83, the music industry was desperate for a new recording medium because the prices and volumes of traditional vinyl records were falling fast. CDs, in contrast, are selling extremely well, and the industry is even trying to raise prices. The music industry fears that new media would confuse customers and ultimately depress sales. And few customers will buy an additional player while the range of pre-recorded material is tiny compared with the range of CDs.

At present, then, both companies are unexpectedly finding themselves left with their convincing product innovations on their hands. Whether the flops will yet turn into hits remains to be seen.

LESSONS FOR PIONEERS

Of course, to start a new core business you need a 'genuine' innovation that creates added customer value. But, as examples from widely different segments show, even excellent innovations will fail if market entry is not timed right. To spot when a 'window of opportunity' will open for which product, its marketer needs to understand how it interacts with its surrounding network infrastructures, software environments and interface standards, as well as whether and how it depends on a critical mass of users. These factors must be included in concrete strategic decision-making if a company wants to prepare successfully for the often unpredictable 'trigger points' for new businesses and obtain the essential positive feedback.

With the fax machine, for example, positive feedback from an increasing number of subscribers is generating a trigger point for dramatic growth. After years of low figures, a critical mass of fax users was evidently reached in Europe around the mid-1980s, and penetration density exploded. A further positive feedback loop lay in the fact that high volumes led to drastic cuts in fax machine prices, stimulating growth still further.

The experience of European suppliers of digital televisions (HDTV) is almost exactly the opposite. They came to grief on the negative feedback of excessive infrastructure requirements, the importance of which was probably underestimated. The European semi-digital television system which was intended to bring about major improvements in sound and image quality required changes not just to the transmission route and TV sets, but also throughout the 'software' producing industry (for example, film studios) and studio equipment.

The transition was to have been made easier by compatibility with the existing PAL standard and by not recoding films. But this was not enough. The new transmission infrastructure via satellite and conversion of the entire hardware base – from the customer's television set to the broadcasting station's studio equipment – beset the Europeans with insurmountable obstacles. Despite billions in European Community subsidies, the innovation could not get established. The fully digital American system might have a better chance. It requires almost the same changes, but its interactive applications offer higher customer value.

What networks are available? Can the existing telecommunications networks be used? What adjustments and extensions are needed? What software can be used? To what extent does new software have to be developed or modified? Is the new development covered by existing standards, or are new standards necessary? What is the balance of political and economic interests? Will standardization efforts be well received? Does the innovation meet with the approval of other participants in the economy, or does it actually depend on their added value? These and similar questions about the environment are often more relevant to assessing a future business than technical considerations (see Figure 6.2).

An accurate understanding of the infrastructure can show a company how likely a product is to be a hit, and what can be done to improve its chances. Once the interdependencies have been properly understood they will rarely cause problems. Much more often they provide additional degrees of freedom that extend the strategic game. After all, positive feedback from the environment can be initiated or intensified by the company itself.

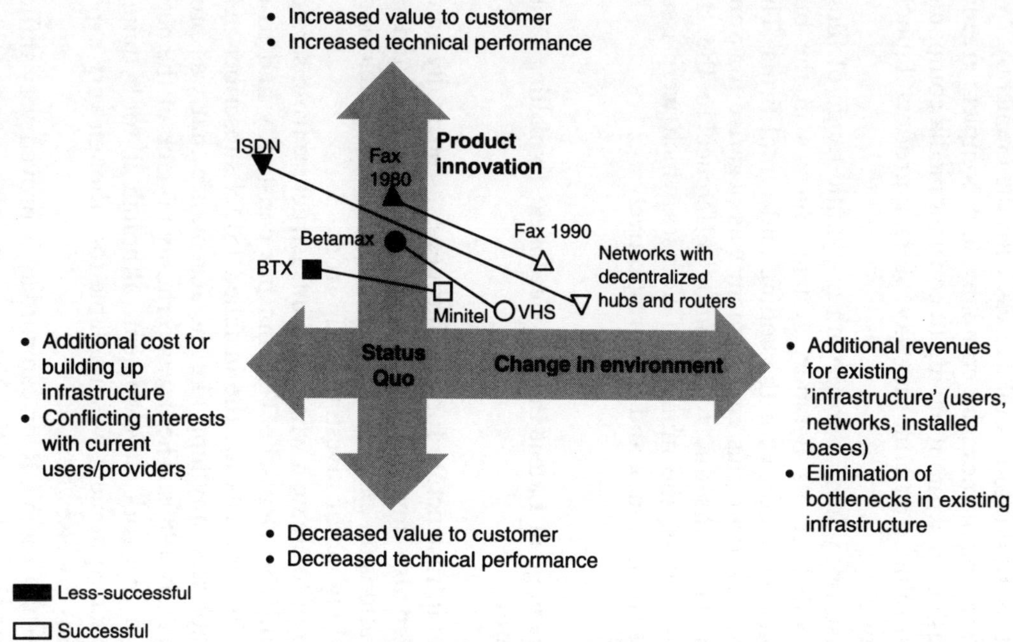

Fig. 6.2 *To become a real success, many product innovations require additional changes to their application environment*

Generating Your Own Positive Feedback: Creating Customer Value Independently

Being in absolute control of your own decisions in a new market is the easiest position to start from. Nintendo turned an independent strategy of this kind into a total success with the Gameboy. With its own hardware and games, the company had complete freedom to design its business – without having to consider media companies, for example. Nothing stood in the way of fast product launch and dissemination.

In contrast, Sony had to cope with a gigantic range of films that had to be coded for its Betamax system. Success in the business depended on interaction with the coding and rental firms. The fact that the feedback from this environment was negative for Sony and positive for JVC was due to a minor difference in the starting conditions – that is, the number of licenses which were issued to third parties for Betamax and VHS respectively.

Taking Advantage of Latent Positive Feedback: Exploiting Existing Infrastructures

Creating the infrastructure for new businesses is normally an expensive and often also time-consuming affair. It is far better and safer to build innovations on existing structures, for instance by closing gaps, without making major investments up front.

The attempt to create a whole new infrastructure from scratch was made with ISDN. Deutsche Telekom has constantly had to drastically revise its forecasts for the installed ISDN subscriber network, which has not seen anything like the acceptance that was predicted for it in the early 1980s. There was certainly no lack of the necessary investment in network development, although it took quite some time. End-user equipment was available too, though very expensive. The problems lay elsewhere.

In contrast to the VCR or camcorder, it proved very difficult to communicate the value of ISDN to the private customer. His main interest continued to be in voice services and he saw little or no use for other facilities – if he knew about them at all. Why should he spend extra money to get them? For business customers, ISDN came at a bad time. Integration of speech and data was not initially high on their list of priorities. When later, in the 1990s, proposals were made to link 'local area networks' (LANs) into 'wide area networks' (WANs), ISDN with its 64-kilobit service was nowhere near adequate. The additional value of ISDN to the business customer could not compensate for the substantial switching costs.

Data technology companies such as CISCO, Wellfleet or Synoptics took a different route. They undertook to solve for their customers the urgent problems arising from PC networking, from different networking standards, and from the need to create the simplest possible links between the various user groups and provide access to the relevant information with ever-increasing data volumes.

The problems are obvious: LAN and WAN network architectures are not compatible; the fixed bandwidths of conventional telecommunications lines are not suited to transferring intermittent groups of data – either bottlenecks or overcapacities occur. In this situation, customers are looking for solutions they can implement quickly, instead of waiting years for the development of a new infrastructure.

CISCO and similar companies offered such solutions – without the superstructure of a total network concept integrating different services and massive infrastructure investments. By focusing on data networks instead of immediate integration of speech and data and with products such as routers and intelligent hubs, they generated enormous cost savings for their customers. At the same time there was very little need to replace equipment.

High customer value without high switching costs or investments in network infrastructures was their success formula, and it brought them annual growth rates of over a hundred per cent in some cases. Major corporations, which generally tried to implement the 'full-scale solution', found only marginal growth in almost every case.

A temporary solution may also be opportune when a company finds itself under severe pressure to generate revenue. In the mid-1990s, this is the position of the future American operators of the PCS digital mobile-phone service. For technical reasons it might make more sense to wait for the new CDMA standard before installing the systems. But many companies cannot wait that long – they need to start collecting fees as soon as possible to cover the high costs of their licences. Several of these licensees are therefore going to install the proven European GSM system to start with, so as not to risk a loss in revenue from operating an untested technology, and will change to the CDMA system later.

For the European GSM suppliers, who are left with very few new markets, this is a very happy turn of events. They have the chance of an additional business without having to invest much more.

Initiating Positive Feedback: Involving Third Parties in Adding Customer Value

Complete independence is very much rarer than most companies would like to believe. They think they can take decisions as free agents, and do not notice the host of other interested parties until they meet with resistance or the hoped-for success fails to materialize.

Nintendo mastered one variant of this risk when it safeguarded its 'independent' Famicom success by tapping in to third parties' software ideas, while wording contracts so as to keep overall control firmly in its own hands. Conversely, it failed in an attempt to repeat its success with a networked Famicom. Famicom proved unable to establish itself as a terminal for interactive games, nor were there enough takers for the home-shopping and home-banking applications.

The company, continuing to bet on independence, had not thoroughly analyzed the success factors of these businesses and had underestimated their dependence on infrastructures. Interactive businesses can only work properly when a positive feedback loop is completed consisting of end-user equipment, network infrastructure and service providers. The drivers in these loops are likely to be the service providers.

If a new business is to be created through positive feedback by mobilizing third parties, the central question is: how should value-added be divided up among the various players? If the share-out disadvantages a major player, that player will hamper the business. If, on the other hand, conditions can be shaped so that the maximum number of participants benefit or expect to benefit, then the business will make rapid progress. Obstacles will be removed and standards which would otherwise take far too long to produce will be completed; compromises will be made which were previously considered impossible.

A common understanding, common interests and the distribution of value-added over a wide spectrum of players will probably help give the American HDTV standard its breakthrough. It has a far better chance than the abandoned European standard. Why should this be? The reason is that the fully-digital US standard is designed for interactive television and not, like the European one, merely to improve sound and picture quality. Broad customer value gives many providers the opportunity to develop new business, and the high infrastructure costs can be widely spread.

Cable TV suppliers and makers of electronic games, suppliers of home shopping services or adult education programmes – everyone wants a share in these new systems. This creates the basis for a positive feedback loop involving a very large number of participants – a sufficiently strong one, perhaps, to overcome the hurdle of infrastructure investments. The ultimate winners in this multimedia business remain to be seen.

SUMMARY

The combination of technological progress that allows specific new applications with an environment that is ready and willing to accept such innovations defines very narrow windows of opportunity for new electronics businesses. Entry conditions are favourable for only a very short time. Dramatic growth rates are possible at a point where there is positive feedback between innovation and environment. If a company misses this golden moment, the market will already have been parcelled out by the time it gets there.

The number one success factor is accurate observation. Where are the new markets? What latent demand is there? What is the infrastructure like? How far advanced are technological developments, and when can the next leap forward be expected? What kinds of positive feedback will there be?

- Successful companies not only know about the speed of innovation, they are also aware of the interaction between their product innovations and infrastructural changes, such as the investment needed to develop a network or the costs of switching to new software. They understand the mechanisms of positive feedback and the resulting, often substantial, growth potentials better than their less-successful competitors.

- Where market forecasts of even reputable analysts vary by a factor of 8 to 10 – as for broad-band ISDN communication technologies – successful companies do not base their product–market strategy on long-term planning. Up to the point of product launch they react extremely flexibly to last-minute changes in demand and infrastructure. When building new infrastructures they implement their own standards as early as possible with the help of broad-based alliances.

The most successful companies are alert and ready to seize the initiative themselves wherever necessary; either by actively creating positive feedback, initiating it by mobilizing third parties, or exploiting existing latent positive feedback.

7 Nothing Fazes the Problem-Solving Organization

If companies were as progressive organizationally as they are technologically, they could achieve far more. As it is, modern control systems that pick up information in direct contact with the grass roots and process it locally are more widespread on the engineering side than in internal organizational structures and procedures. There, information and decisions still get tangled up in a variety of filters on their way through official channels. But in the fast-moving electronics industry, the companies that have realized the ideal of self-management most completely are the most successful ones.

The driver of a state-of-the-art car is free to concentrate on his 'strategic' objectives – deciding on the route and responding to the traffic situation. He is largely relieved of the 'operational' functions because, for example, antilocking and anti-skid systems free him from the need for fine control – particularly in critical situations, such as emergency braking or acceleration on slippery roads. That is handled by the parts of the car in direct contact with the road. Wheel-speed sensors in the wheels, anti-collision sensors in bumpers and pressure sensors in tyres collect the information at the 'grass roots', where most of it is also processed. All-wheel drive, all-wheel steering and active suspension control ensure that all resources that will benefit safe implementation of rapid changes are mobilized.

Traditional corporate management, in contrast, still prefers to take too many decisions at the corporate centre or at middle-management level, and suffers from filtered information and long decision paths. Nowhere is there a more vital need to improve this situation than in the electronics industry. The superior company here knows how to use scale effects internally, improves product and process technology as fast as possible, and draws on worldwide resources for R&D, production and sales more skilfully than others.

Quantum-leap innovation and continuous improvement alternate here, driven by strong competitive pressure. And that makes supreme demands on the efficiency and adaptability of the organization, through all stages of the value chain and the product life cycle.

Product development has to integrate new customer wishes and technologies. It is called upon – particularly in the field of component technology and the large systems and industrial electronics/measuring systems segments – for great innovative achievements. Marketing and sales have to get to the market fast to ensure that market potential is exploited worldwide at existing and new customers: opportunities to earn high margins and thus recover the cost of innovation are only available at the start of a life cycle. After product launch and as market penetration develops, rapid continuous improvement is more important – for example, to improve yield in chip production (that is, the number of good chips per silicon wafer) or reduce the insertion and soldering defects in printed-circuit-board manufacture by fine-optimizing the automatic equipment and improving the board layout. The purpose is continuous improvement of processes, from manufacturing to all the 'indirect' tasks such as materials provision, servicing and maintenance.

The winner here is the company which has the best mastery of problem-solving 'on site', as a rule at the organization's lowest operational level, and which combines these problem-solving skills with intelligent strategic positioning and responsiveness throughout the organization. These skills are most likely to be found in small, innovative, highly dynamic companies. They can exploit their natural advantages particularly well in the electronics industry, since, in contrast to metalworking or chemicals, the small companies are not held back here by the need for high capitalization and large plant.

But the time of the problem-solving, fast-learning, self-managing organization has evidently come, and it is finding its way into more and more large companies. This is demonstrated by the widely-observed experience of ABB (Asea Brown Boveri) and General Electric, as well as other particularly successful competitors. But what the small firms gain by evolution, the large companies have to acquire by a generally painful cultural revolution. The most important features of this organization are fundamentally different from many traditionally unshakable concepts:

- Small and agile – not large and lumbering;
- Integration – not functional separation;
- Entrepreneurship – not command and control structures;
- 'Try it!' – not 'Don't experiment!'

SMALL AND AGILE – NOT LARGE AND LUMBERING

Agility and fast responsiveness are, of course, directly related to the purely formal size of the operating units. In the problem-solving, self-managing organization, these are relatively small and transparent. But they only become really agile when endowed with a high degree of individual accountability.

Small, Transparent Units

The call for short paths, clear assignments, direct communication between employees who know each other personally, and not least closeness to the customer, comes as no surprise these days. Many companies have already acted on the implications and are breaking up their large organizations into more manageable business units.

'Small is beautiful' seems to be the undisputed rule. Just how small needs to be decided from case to case. The optimum size will change, driven by various factors over time in the individual segments. In consumer electronics, the critical mass for a business unit is increasingly determined by rising advertising spend. The need for expensive media and promotional vehicles such as television and sports stars and the trend towards globalization mean that consumer electronics depends on real 'hits'. More complex products usually require a minimum size of company to secure the payback on investments in equipment for development and manufacturing. In major telecommunications installations or with complex medical electronics (for example, computer tomographs), where it is not so easy to obtain standard modules from outside contractors, the original equipment manufacturer has to have the minimum size on its own, and thus has to realize a minimum share of the market.

In the computer industry, particularly in the case of PCs with their standard modules, the original equipment manufacturer may be smaller. The module suppliers themselves bear a part of the investment. Here, the OEM has to gear its minimum size to the logistics and assembly input needed. In the still largely problem-solving-based business of industrial electronics and measuring systems, it is usually only necessary to reach a critical mass of know-how, which will be self-sustaining.

Changes in the nature of the business – such as the rise of mass products like programmable logic controllers in industrial electronics or the emergence of standard modules in telecommunications – may also change the ideal size. Successful companies anticipate these changes and stay close to the economic minimum size for a particular activity.

In consumer electronics and computers/communications, the improvement champions' business units are only about half the size of those at their slower colleagues; in the large systems segment, the fast adapters' business units are 30 per cent smaller on average. Industrial electronics generally manages to operate with small business units irrespective of whether the companies are fast or slow adapters, their units never being more than half the size of those in other segments, and generally very much smaller, as shown in Figure 7.1.

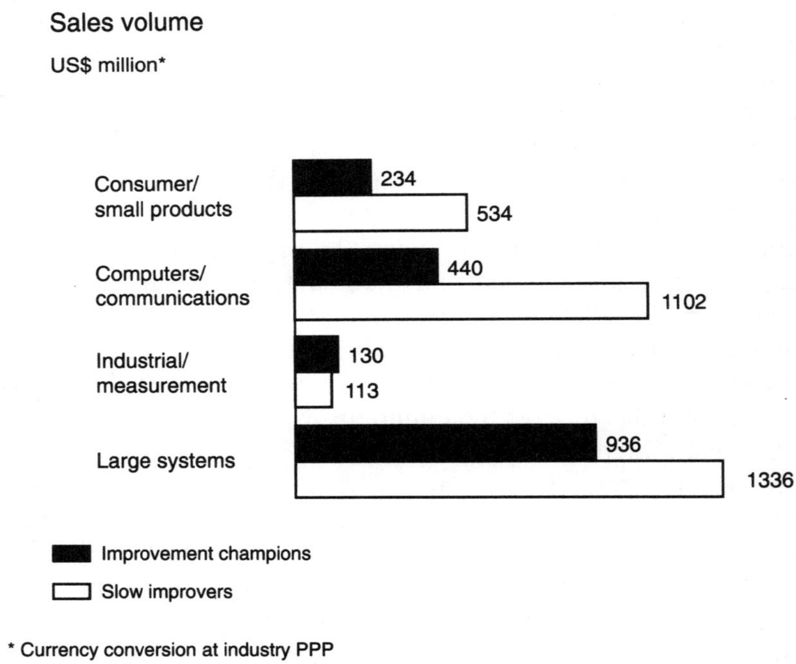

Fig. 7.1 *Small units*

Small units are so much more efficient largely because of their inbuilt transparency, and because they depend much more directly on market success. This latter feature is particularly marked if a unit is tailored to a particular customer or product segment.

ABB's 'radical reform' showed in extreme form how far you can go in this direction. With its 5 000 or so legally-independent, sharply-focused profit centres, ABB manages to enter practically every market in all important segments as a local competitor, and to establish close contact with the customer. Whether such extreme fragmentation can succeed in the long run or will overshoot the mark is something that only time will tell.

Smallness pervades successful companies right down to the 'operative level' of the teams that perform the day-to-day operational work in the line and project organization. As the long-term study shows, small teams with eight to twelve members are most productive. They have been proved to achieve better project results – they arrive at solutions faster since the membership and the distribution of tasks are of manageable dimensions.

Successful electronics companies take these findings into account in their planning. In the segments of consumer electronics, computers/communications, and industrial electronics/measuring systems, companies keep their development teams 20 to 25 per cent smaller than those of less-successful competitors. In large systems such as mainframes and large medical electronics systems this difference is even more pronounced, at 70 per cent.

How extreme the impact of introducing lean structures is, particularly on the staffing of top management levels, is shown by a simple example. In one traditional factory about 1000 line employees were managed by 150 management personnel. The management organization had six levels, from the works manager downwards. Now, two years later, about 760 line employees are producing the same output in a lean, optimized factory with 75 managers and half the hierarchical levels. The number of factory employees has consequently gone down by a total of 27 per cent – three-quarters of the line employees have remained, but only half the managers (see Figure 7.2).

But this particular example also shows that the creation of small units alone is not enough. These units will only be as mobile and capable as the company wants them to be if the connection between performance and economic success is clearly visible to them – and if they can influence them under their own independent responsibility.

CASE EXAMPLE

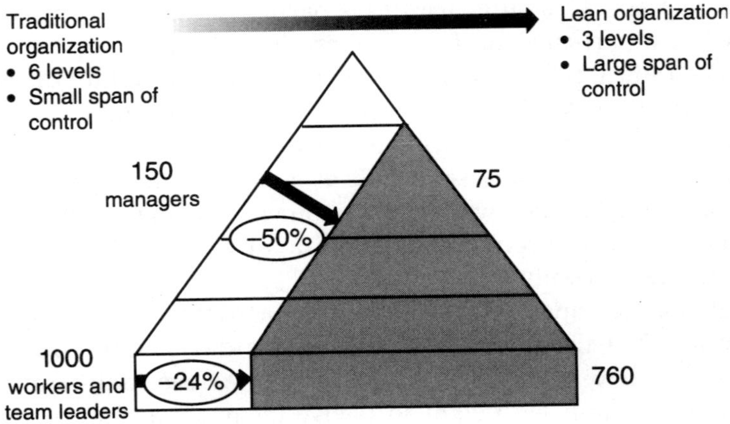

Fig. 7.2 *Achieving lean, problem-solving organizations*

Mobility Through Independent Responsibility

The improvement champions in all four electronics segments predominantly run their business units as 'medium-sized' profit centres as shown in Figure 7.3. Extensive powers of self-management are not, however, the sole province of the business divisions. The concept of self-direction is in place right down to the lowest hierarchical level.

Japanese and American companies have shown the way at home, delegating problem-solving responsibility as far down the line as possible and transferring a lot of tasks from foremen and group leaders to self-managing teams (line 'empowerment'). Those who believed that this idea would not survive the journey from the Far East or the USA have long since found otherwise. Japanese and, lately, an increasing number of American companies operating in Germany and other European countries, have been as successful with these concepts in Europe as at home – and more and more of their European colleagues are following their lead.

The array of tasks assigned to the 'line' and to 'teams' has grown markedly since the principle of delegation was introduced. Once almost entirely restricted to pure machine operation, their standard repertoire now often includes quality-data capture and quality control, material call-off, maintenance, and the constant search for ways to improve productivity.

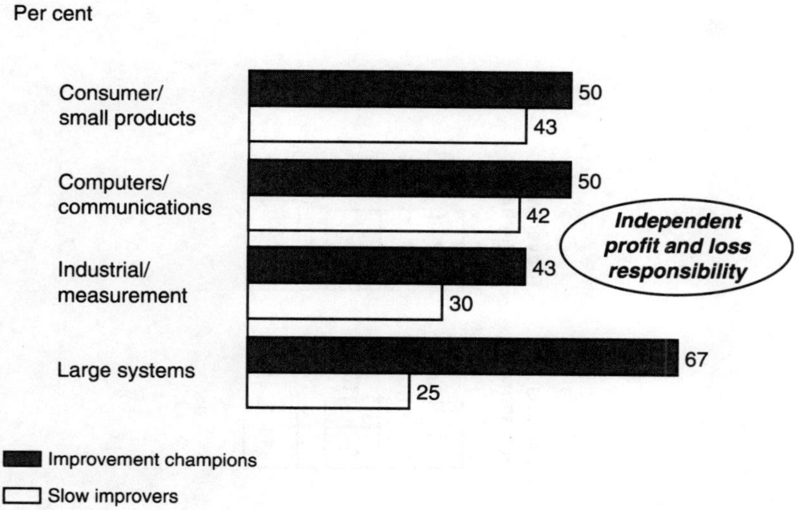

Fig. 7.3 *Production facilities run as profit centres*

Japanese manufacturers traditionally lead the field, but by 1994 European and American manufacturers had caught up with them in the 'level of delegation', as shown in Figure 7.4. But there is still a lot of hard work ahead before comparable results in 'empowerment' will be seen at European and American companies.

One European electronics manufacturer runs a main plant that can already be considered as a model, so successfully did it implement a flat hierarchy. One works manager and three foremen run the 360–man factory, operating in three shifts (about 150, 150 and 60 workers for the early, late and night shifts respectively). Two production engineers are available to the works manager for project assignments. Each of the three foremen oversees one shift, and the two day shifts are additionally run by three supervisors each. If a foreman is unavailable, his deputy is one of the two production engineers.

Creativity and commitment, otherwise – in many people's opinion – only experienced by employees in private leisure activities, flourish readily at shop-floor level in this environment, and are also readily welcomed. Gradually, team members have been entrusted with new and more and more challenging assignments, resulting in the elimination of some support functions. One example is the maintenance function, which relinquished its responsibility to the 'line' some time ago; quality control will soon be overtaken by the same fate.

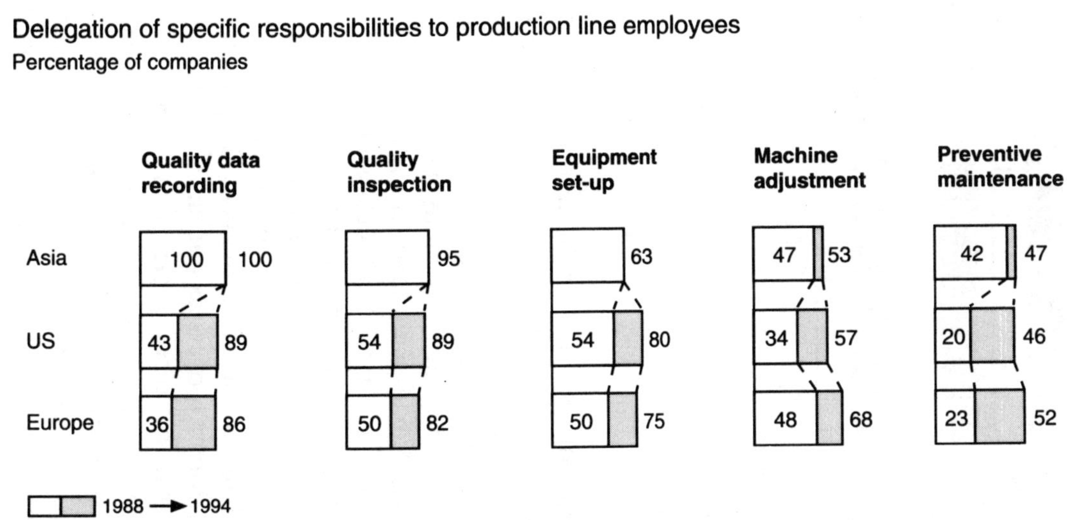

Fig. 7.4 *European companies have delegated less responsibility to shop-floor workers*

There are still managers who consider such lean management structures and lack of supervision risky to the point of adventurousness, but all the experience suggests that the deterministic-dirigiste approach is ultimately more risky, because it develops improvement potential too slowly and leaves too many gaps.

In one case, a third of the defect rate for inserting chips at a telecommunications factory was due to the injection-moulded lugs of the chips catching on each other when picked from the feed channel by the automatic inserters. This sort of problem can only be recognized and quickly and permanently resolved by a production-line worker – in this case by angling the feed channel.

Self-managing teams can be used in many areas besides manufacturing, however. Administration and sales are two other functions where forward-looking companies have advanced from tentative initial experiments with delegation, such as having staff arrange vacation schedules independently, to generally comprehensive self-management projects with more and more demanding responsibilities.

Highly-skilled working groups are thus created. When processing orders, for instance, groups in the sales function not only manage volume smoothing themselves, but also take active steps to secure better coordination between field and office staff in customer care and to reorganize inefficient procedures in order processing. And finally, one measure increasing productivity for the company and flexibility for the employees is increased independence in the planning and flexibility of working hours.

INTEGRATION – NOT FUNCTIONAL SEPARATION

The second key to creating a problem-solving, rapidly-responsive, and fast-learning organization is the integration of information flows and processes – involving all internal functions, but also important suppliers and customers (as discussed in Chapter 3).

Internal: Cross-Functional Process Chains

Successful companies have recently been placing much less emphasis on functional divisions and specializations. The best ones consider the benefits of combining individual corporate functions more valuable than the scale and learning-curve effects derived from specialization. (They are exploiting similar effects more by focusing on customer groups and selected product/service packages and product ranges.) This applies to the business units just as much as to lower-level teams, which they set up with all the functions necessary for successful operation.

In the business units, successful companies build their structures around a small number of core processes – such as the process of creating new products from the original idea up to product launch; the total sales process from the first customer contact to completion of any necessary installation and training work; or the maintenance process from the collection of information (on defects) in the field through to repairs, servicing and possibly design changes. To do this, they create more or less complete process chains with end-product responsibility and with the simplest possible inside-out interfaces.

Companies with traditional organizational structures find it difficult to change over to integrated responsibility for the end product. Even when restructuring, they often leave important core-process stages or key responsibilities in different organizational units. But reluctance to make a clean sweep may endanger the success of the restructuring operation.

Many companies have now introduced 'product management' for the product-generation and product-family-updating processes. All too often this coordinating function fails to achieve the desired effect, and proves a paper tiger. If, for example, research and development keeps its sovereignty over the allocation of the development budget and responsibility for its employees' career progress, it may develop a distinct life of its own just like before. It will tend to plan its development portfolio solely around technical priorities and personal preferences. Market needs are neglected because they are often not perceived; wrong investments, dangerous market gaps or expensive corrections after product launch are results that may cost the company dear.

Keeping the old sales function is just as much of a problem; this can easily lead to equal treatment of all customers regardless of the segment-specific differences in product offering, customer wishes and competition. Companies do themselves no favours with this approach, as one European supplier of telecommunications systems learned. For many years its order processing department made little or no distinction between small businesses and major companies. Simple telephone systems with a few lines were given much the same care and attention as large ones with 1000 and more connections. This meant that unacceptable delivery times were imposed on the customers. The supplier itself was penalized by inflated handling costs and large, unintended gaps in the coverage of important customer segments.

Reluctant redesigners are still trying to solve problems such as these with liaison functions, coordinators and monitoring departments. Unfortunately, however, after years of having these functions in place, after they have built up a staff of their own and consumed substantial budgets, it turns out that instead of solving the problems, these functions have often helped them to get more firmly entrenched than ever.

This is because these institutionalized problem-solvers often do not understand and are not close to the actual problem, and are not accepted by those really concerned. For example, the product planners in one German electronics manufacturer's product-management function found that they were not capable of coordinating development, sales and service. Development continued to take autonomous decisions as to what should be developed when and how, and sales continued to furnish itself with urgently needed items from outside suppliers.

These companies generally do not change for the better until those concerned join forces to solve their problems. This is where teams and task forces, formed *ad hoc* and regardless of organizational and hierarchical boundaries, play a key role. If they have clear support from top management – and their activities get the requisite high priority – if they are largely freed from the organization's bureaucratic pressures, and if they are focused on specific end products with clear objectives, they can release enormous energy in an amazingly short time and handle even the biggest problems. Such task forces then form the nucleus of the new process-oriented structure.

Suppliers: From Information Exchange to Joint Ventures

Successful companies involve their suppliers in their value-added process instead of playing them off against one another.

Both sides profit from intensive, generally longer-term cooperation with a small group of suppliers. Development and manufacturing are in close contact with their 'outside colleagues'. They make sure that suppliers have an undistorted view of their technical and economic expectations. But they also give them the freedom to introduce their own ideas and mobilize their problem-solving potential to the benefit of both parties.

The sort of potential that often lies fallow here is illustrated by the example of a German system producer who was under massive cost pressure and wanted to buy modules more cheaply. The offers he got from the Far East were about 40 per cent below the previous manufacturing costs, due to the low wage level. But when a Chinese supplier dangled the prospect of reducing costs by a further 35 per cent by adapting the design to suit Chinese manufacturing facilities, he really sat up and paid attention. After initial resistance and careful examination, the design was revised, reducing the offer price of DM44 by another DM15. Instead of costing DM80 as it had done, a motor of equal quality now cost only DM29 to produce. This provided the manufacturer with considerable savings which he could use to improve his profit margin or his market share.

As the 'Excellence in Electronics' study demonstrates, integrating suppliers is good for more than one-off improvements; seen as a longer-term prospect, it almost invariably leads to better results in unit cost, delivery times and delivery quality. Japanese and Taiwanese companies have long exploited this advantage. When they receive offers in a new tender procedure they make their choice not only on the basis of the price quoted, but after comparing differences in cost structures and the prospects for long-term cooperation. They then work on establishing a close relationship with the supplier of their choice, integrating the supplier both into the product development process and into manufacturing planning and execution.

However, these companies are also aware of the pitfalls of an intensive relationship with one supplier. It may be a trap for those who believe that their good understanding with their partner allows them to take things for granted. This applies particularly when fresh problems arise as new products, process technologies and materials are introduced. On the assumption that each side knows what the other can and will do, they often do not see new problems until far too late, when making corrections becomes complicated and expensive and the original time schedules are ruined.

This was the painful experience of one medical electronics maker, when the integration of sourced modules led to costly delays in the transition from analogue to digital signalling. Part of the cause was poor coordination of the protocols between the system modules and problems in further processing of the measurements. On the basis of years of collabouration, the parties had very largely stuck to an *ad hoc* approach at the operational level, so that although the nature of the problems had been recognized and discussed during the preparations, the explicit solution was implemented too late.

It is thus advisable to apply the same basic principles as for internal integration and keep supplier relations on a largely objective basis, bringing in new suppliers and purchasing staff from time to time. And finally, an exchange of employees between partners also helps develop the relationship further in the right direction.

With increasing market dynamism and growing complexity, more and more companies are relying on skilled suppliers and partners so that they themselves can concentrate on those areas where they have special in-house strengths. This is particularly evident in the computer industry, where specialist suppliers of components with standard interfaces such as diskettes and hard-disk drives, picture tubes, or LCD displays are gaining the upper hand and the share of outsourcing is steadily mounting. In industrial electronics, where new developments are getting more and more expensive, many companies have decided on joint ventures to bundle resources and spread risks. They can then take advantage of each other's know-how.

It is also important to build and maintain functioning local supply structures at world-class level as a complement to global purchasing of commodity components. The local supply structure will change in the process. To eliminate complexity from supplier relations, suppliers of systems have to be developed; these, in turn, will draw on efficient suppliers of modules and parts.

ENTREPRENEURSHIP – NOT COMMAND AND CONTROL STRUCTURES

Changes scare many employees. The flatter hierarchies that give them new rights but also greater responsibilities cause anxiety. Others see the new situation as their big opportunity, and a number of dormant talents come to light. This was the case with one foreman who, as leader of a manufacturing quality project group, reported to a top-level committee, hesitantly at first, but with increasing freedom, about several-fold reductions and their impact on the defect rate. Asked by the CEO – who perhaps should have known better – why he hadn't always done it that way, he replied: 'Nobody ever asked me'.

Neither employer nor employee can really afford to let skills lie dormant like that. The employees' ideas and commitment are lost to the companies, and the individual employee sells himself short on a strained labour market – and also forgoes added job satisfaction.

To start with, of course, those concerned will have to move out of their comfort zone: managers have to surrender 'feudal' rights, particularly monopolies on information, and workers can no longer be sure they will spend their whole working life gainfully employed in the job they originally learned. To the contrary, they will be expected to bring highly-developed, up-to-date technical skills to the party, combined with the prudence, mobility and risk acceptance of an entrepreneur. And they will be expected to make the development of these technical and entrepreneurial skills their own personal concern.

It is a fundamental shift. Analyses in the US have revealed that employees in traditional large companies who only give 30 per cent of their performance capabilities to their work may well be among the respected and upwardly-career-mobile pillars of their company. Whether or not it proves possible to mobilize as much as possible of the remaining potential and attune reluctant employees to the new course is crucial to the success or failure of the company in not a few cases. Perhaps the greatest strength of the self-managing, problem-solving organization is that it makes it easier to handle such changes of direction. The keys are continuous further development of the staff and a new role for management.

Continuous Staff Development

Personal initiatives are only taken by workers who feel in control of their situation. That is why successful electronics companies aim for their employees to have an excellent knowledge base in their immediate specialist area, in management issues and in social skills. They hire the best educated people possible at all levels, and take more trouble over their further training. Training in these companies comes from the best sources available, and is not to be considered a leisure activity with entertainment value.

Good training programmes are directly related to practice. Training on the job, or dedicated training modules alongside work, ensure that this requirement is met. An excessive time lag between training and application must be avoided. PC training becomes ineffective if it is carried out even as little as four weeks too soon because the computers are installed late at the workplace.

Job rotation has proved to be a particularly effective form of training. Employees at all levels of a self-managing, problem-solving organization take part in these exchange programmes. Successful companies in computers/communications make 30 per cent more use of them than their less-successful competitors, while they are actually twice as popular in consumer and industrial electronics (see Figure 7.5). With an average stint of eight to twelve months in another function, these 'rotating' employees gain valuable insights into what is required of their own jobs, and broaden their information network with new contacts within the company.

In job rotation, too, detail is important. When development engineers move to manufacturing and then back to development, for example, they take with them concrete knowledge about the manufacturability of products which will be used in future product design. Successful Japanese electronics companies continue to set the standards for effective job-rotation programmes, with regard both to the number of employees taking part and to programme content.

New Role for Management

Even the staff of a self-managing organization needs management. It is no longer important, however, to give instructions for everything down to the last detail. Instead, the right general guidelines have to be set and support for teamwork put in place. Management, freed from many operational responsibilities, can concentrate on this task more easily than in traditional organizations. Middle management is transformed from commander to coach. It does more to eliminate obstacles in the organization to make it easier for teams to solve technical or functional problems, and concentrates on personnel development – or, rather, personality development. 'Care' and intensive coaching are a standard component of the manager's responsibilities. And this is far from being a socio-romantic exchange of nice noises. It involves both discussion of the employee's personal worries and interests and clear, objective feedback about his skills and performance, together with an agreement on an individual further training programme.

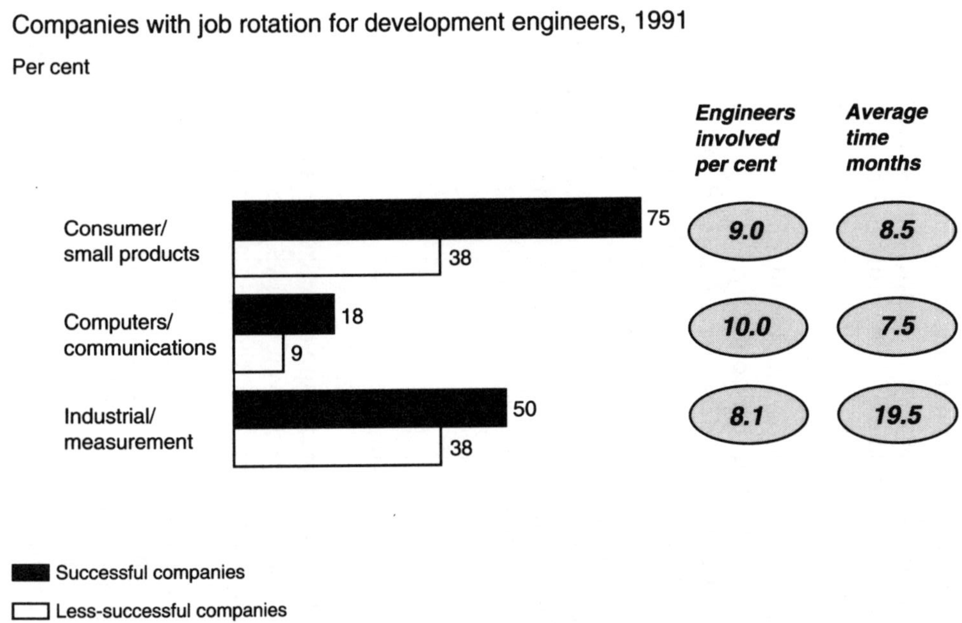

Fig. 7.5 *Successful companies provide more opportunities for job rotation*

The importance of 'caring' to personal effectiveness – even to the employee's state of health – became apparent in one European electronics company. When 10 per cent of the workforce was off sick, the foremen and works manager started to hold personal interviews with those involved. There were no threats and no imposition of sanctions, but the sick rate dropped to 2 per cent – only to shoot up to 8 per cent once the interviews stopped. These interviews have now been made a regular fixture, and the sick rate is down again.

This has nothing to do with being 'soft'. Quite the contrary, in the most successful companies managers at all levels are very well aware that pursuing half-hearted goals will get them nowhere and that the results are likely to lag behind even revised expectations. If problems are not eliminated and performance is not demonstrably increased, there is no way to hide it from employees. If management then tries to sell stagnation as success, resignation soon sets in.

The sustainably successful electronics company therefore deliberately sets ambitious goals, generally derived from internal and external comparisons. In view of the industry's particular dynamics, companies prefer to set improvement goals dynamically in the form of half-life values. This shows how much time should be taken to halve the defect and cost level, for example, or what innovations should come onto the market at what time. Improvement champions compare themselves with the best – not to catch up with them, but to overtake them. They try to reduce processes and products to their core contents, core times and core costs – the theoretical minimum.

And it is also taken for granted that laurels are not for resting on. There are still problems to be solved and new areas worth exploring, so that no limits are set to innovation or productivity improvements. In printed-circuit-board manufacture, for example, the minor disruptions caused by stockouts are only noticeable and relevant when the defect rate is low – component delivery then becomes the most critical problem. Once a solution has been found, a new problem rises to the top of the priority list, and a further performance leap becomes possible.

Self-managing teams in successful companies are continually measuring their goal attainment and how their own performance has developed. Objectives are therefore guides and goads in one – and not just a controlling tool. Employees should not feel overwhelmed or pressurized, but should feel that their problem-solving skills are constantly being developed and their self-esteem bolstered.

'TRY IT!' – NOT 'DON'T EXPERIMENT!'

Performance, motivation and personal responsibility will only develop if the corporate signals are right. Delegation of responsibility and tasks to self-managing teams must be accompanied, therefore, by changes in the evaluation and reward systems.

To start with, team performance has to form a significant component of individual assessments. Successful companies put their faith in a balanced mix of individual and team aspects. They are aware that the emphasis must be on reward and not punishment, but they do apply sanctions stringently in the few cases where they are really necessary.

As a result, these companies ensure control and stimulus within the teams on the one hand, while taking care that the stronger members of the group help the weaker ones on the other. The net result is that team performance is actually significantly better than the sum of individual efforts. In one European electronics company, the productivity of a manufacturing line rose by 27 per cent simply as a result of the change from hourly pay to group performance pay.

However, the principal distinguishing feature of the self-managing, problem-solving organization is that its choice of motivators and the type of achievements rewarded signal top priority for improvement initiatives.

Positive Motivators with Clear Transparency of Performance

Successful companies rely almost exclusively on positive motivation; they have more faith than others in financial incentives, career advancement, and appointments to ambitious projects and tasks.

Tangible motivators have the unpleasant characteristic of rapidly losing their effect and constantly stimulating an appetite for more. Goal conflict between the growing demands of employees and the necessity of recouping rising costs with increased productivity is built in from the start. Extended responsibilities, on the other hand, are more likely to lead to an improvement in the employee's skills and abilities. His self-esteem is raised and productivity increases to match; goal conflict can be avoided.

It is important, therefore, to make sure that performance is rewarded by extended responsibilities and enhanced decision-making powers in addition to money, praise and the insignia of rank.

Successful electronics companies put this into practice in development; for example, by making much more frequent use of status-enhancing and image-promoting measures to reward successful employees, such as public recognition and assignment to attractive projects.

To stay on the crest of the wave, to ensure that the company never ceases to be surprised by the surprising results of creativity, and to maintain a constant speed of change, internal competitions should be staged regularly. These might take the form of controlled competitions between areas with product and profit responsibility, or between different manufacturing teams. Little merit has been found in competition between functions and individuals; it may even be counterproductive, because in almost every case it leads to people defending a personal advantage by withholding information or hampering processes.

Draconian penalties such as dismissal or salary reductions if goals are not attained are the exception in successful companies, and are reserved for those who jeopardize corporate performance by an obstructive attitude and persistent misbehaviour. Obstructive and blocking behaviour will not be tolerated in any successful company; the culprits are invariably removed from their posts – regardless of whether they are at the top of the organization, in middle management or on the production line. A successful turnaround often begins with the replacement of the CEO or General Manager and major sections of management.

Don't be Afraid to Experiment

Isolated failures carry no sanctions in successful companies. When new technological solutions and new applications and products are being sought in upstream development, when new customer groups are being won in sales, or a new regional market is to be entered, employees must leave the beaten track and think of new ideas, suggest them, then try them. In a culture intolerant of mistakes, new approaches are never adopted because the personal risk of failure is too great for each individual development engineer or salesman. The result is a massive loss in innovativeness, self-management and flexibility.

Much the same applies to continuous, efficiency-oriented improvements in day-to-day business. In extreme cases, an intolerant environment frustrates the introduction of even the smallest improvement ideas by line employees. The play-it-safe mentality predominates. The source of further development dries up: for the individual, the proven status quo carries less of an immediate risk than a change where the consequences are uncertain.

Analyzing his manufacturing processes, an American chip producer found that the machine operators used five to six times more photosensitive resist on the wafers than was actually technically needed. They were doing it to forestall the problems that often arose after lithography. Over and over again, the photoresist layer proved to be too thin; line workers were blamed for not being able to follow the development engineers' instructions properly. 'We developed it and calculated it so it would work, and you couldn't get it right' was what they were hearing. To avoid sanctions, they built in an enormous safety buffer and wasted huge potentials.

The squandering of resources did not stop at higher resist consumption; it went much further. Five to six times as much resist requires larger stocks in production and in the warehouse. These larger amounts must be more frequently ordered and delivered, tested, put in stock and drawn out again. They lead to inventory, space and handling costs. Consequently, the entire in-bound logistics chain is affected, and has to spend five to six times as much to satisfy the personal security needs of line employees. The excessive safety mentality also led in this case to further safety margins being built in at every stage of the logistics chain on the principle of 'everyone should have his own stocks', so that the inventory on the production floor was enough to last three months instead of three weeks.

Safety margins like these may, in sum, increase the capital and manpower tied up in inventories and process capacities by a factor of three to five, and significantly limit agility. Such buffer stocks can only be reduced or avoided by a solution-oriented, objective dialog between all parties concerned, without trying to find someone to blame.

A risk-tolerant, problem-solving orientation is particularly important where large-scale innovation and high growth rates are concerned. The basic rule is: the higher the innovative content of the activity and the product, the more willing you must be to take risks.

The differences in risk levels also apply between individual functions, continuing down to the level of the individual employee. The upstream stages of development and production engineering have to be more innovative than product development or manufacturing and assembly. Similarly, looking after established core customers entails less risk than acquiring new customer groups or opening up new markets.

SUMMARY

Improvement champions generally have a problem-solving, flexible organization. Mechanisms reminiscent of the practices used by small companies also give large corporations the necessary flexibility and action capabilities to achieve superior performance in restructuring and growth:

- A new organizational design puts in place small, transparent units, which should if possible have profit-centre responsibility for the success of a business. These units are organized along the core processes of the business with extremely highly-integrated information flows and procedures both between internal functions and with customers and suppliers.
- The ability of all employees to solve problems independently is systematically cultivated by delegating responsibility, building skills, and establishing a corporate culture tolerant of error. Many 'experts' on the lowest hierarchical level turn into 'entrepreneurs'.

For the companies that stay at the top, the self-managing, problem-solving organization is an important reason for their success – even the highest demands in restructuring and growth cannot faze them. Chapter 8 shows how, given the necessary determination at top-management level, the foundations of such an organization can be put in place within six months. This can only be good news to other competitors, especially for the many, particularly European, manufacturers facing an uphill climb back into the top league.

8 A Catch-Up Strategy Must Combine Restructuring and Growth: From Shock to Stimulus

Decisions on radical improvement initiatives are easier if there's nowhere to go but up. When the idea of closing a business or moving to Asia starts to get close to home, any option for continuing at the home location suddenly becomes open to constructive discussion. But even for profitable European companies, the performance gap to the world-class is often uncomfortably wide. The need to catch up is urgent. It is also a painful process, with the best increasing their lead every year, and the yardsticks for restructuring and growth getting harder to satisfy all the time.

'Some find customers to pay for their high costs, others don't.' This is how a director of one electronics company described the difference between profitable and unprofitable European high-tech companies, thus confirming a result of the 'Excellence in Electronics' study: in Europe (and nowhere else) there is no clear correlation between productivity and profit in the companies surveyed.

Cost disadvantages of 30 to 50 per cent, like those separating the world's unsuccessful electronics suppliers from the successful ones, are a dangerous affair even if the financial crunch has not yet come. And even if the bottom half of the German product range has not yet had to be scrapped, and outsourcing negotiations are not yet under way with Hitachi, Mitsubishi, Matsushita, Daewoo or others.

Reducing costs by doubling worker productivity seems a herculean task to many management teams, and they are not wrong. But it does not stop there. World-class companies increase their productivity by around 7 per cent each year, which implies that in the 2 to 4 years normally scheduled for the catch-up programme, the gap will continue to widen. In parallel to the cost competition, therefore, innovation and regional expansion would have be used to double or treble the volume growth rate.

The situation is not unusual: a company that has been losing 1 per cent of return on sales in a DM4 billion business every year for several years in succession has reached the break-even point, and is now threatened with losses of DM100 million, possibly more as prices plummet. The company's total cost disadvantage of 25 to 30 per cent in comparison with world-class suppliers is likely to go on increasing, in view of their annual productivity improvements of 6 to 8 per cent.

Is it time to throw in the towel? Only if discussion on the catch-up programme stalls at the elimination of the DM100 million loss, or if a cost reduction of DM300, DM400 or even DM500 million at constant volume is all that is considered. If, that is, the company is not prepared to face the fact that its product and process costs are currently at least DM1 billion too high for world-class productivity, and that far greater economies – of the order of DM1.5 to DM1.7 billion – will have to be made within four years.

This is often coupled with a number of other problems: expansion into regional growth markets is impossible because costs are too high for low-price countries; the company cuts back on R&D instead of attacking markets via innovation; and it withdraws from low-cost segments, further exacerbating its overall position.

Of course, at first sight the escape route does not seem particularly enticing either. If manufacturable design reduces value-added in production by 25 to 30 per cent, process improvements to world-class level contribute to 50 per cent overcapacity, and corresponding growth is unattainable, the only remaining option is to downsize manufacturing structures. Painful as these cuts may be, production staff and management are usually ready to cooperate. Intervention in other areas seems harder: halve all indirect functions, improve sales productivity by 50 to 100 per cent, and double R&D productivity – with today's IT systems, today's people? Suppliers who will go along with massive cost reductions? 'Unthinkable!' is often the first reaction.

And the mere thought of the employment situation: 5 per cent fluctuation a year will not absorb the adjustment needed; redundancy plans – and in some cases even plant closures – are unavoidable. Is all this justified in a 'reasonably sound' business? The break with earlier positive messages, the discussions with works councils, defending previous actions to the supervisory board – it all mounts up to form an insurmountable pile of problems and hassle.

And yet, more and more companies are reluctant just to capitulate in the face of superior competition. Management and labour – manufacturing managers, development engineers, and team members who have run initial benchmark tests and know what world-class standards are – go on the world-class attack. The profit from reaching benchmarks will allow price reductions, thus making customers the principal beneficiaries of world-class programmes. Resources will be available for investments, too – for product substitution at shorter intervals, for higher-performance low-end products, and for the establishment of completely new businesses. And funds will also be available for capturing new markets, maybe in neighbouring European countries, some previously inaccessible because of their lower prices, or for entering markets in the Far East or developing sales organizations in Eastern Europe.

These companies are entering an era of enormous challenges and undreamed-of opportunities. They know that if they focus all their attention on the necessary redesign programmes success will soon follow. Practically every employee must be involved here, however, because the whole company is being turned inside out, all areas minutely searched, all actions challenged. The top management team will keep a tight grip on the reins during the uneasy times of the catch-up process. Its creativity and direct participation ensure that the whole programme, generally lasting 3 to 4 years, clears its many hurdles and that increased competitiveness manifests itself as early as possible (see Figure 8.1).

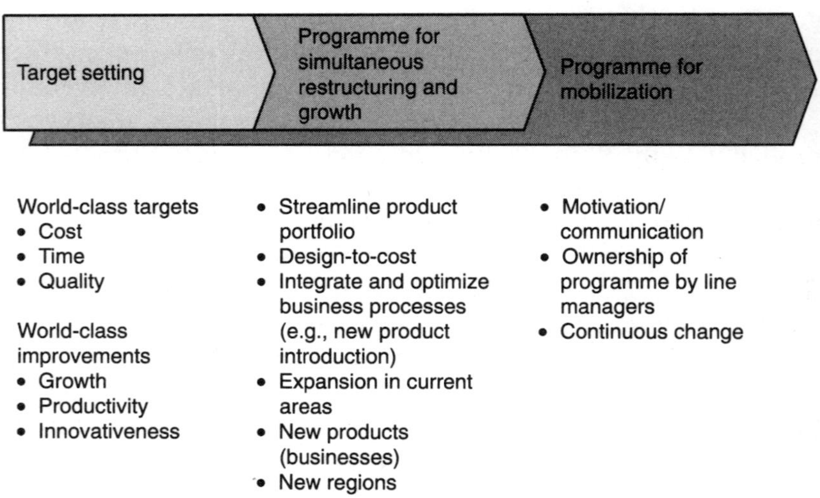

Fig. 8.1 *Answer: total redesign to world-class*

- Top management takes care to establish a sufficiently ambitious, but also well-founded, *set of objectives*. It sympathizes with employees' reactions to the shock, having made a point of informing them frankly about the size of the gap to be closed, and channels employee concerns into a positive attitude of 'if others can do it, so can we'.
- Top management plays a hands-on role in *developing the overall programme*. It allows no doubt to arise that the goals have to be met – in full. However, it also generates enthusiasm across the whole company by creating the appropriate organizational framework and freeing business units from restrictions.
- 'Sacred cows' are deliberately challenged and risks taken during *restructuring,* but new *growth* is pursued just as aggressively, and ideas for new businesses and products are encouraged.
- Throughout the programme, *visible support for project management* is very high on the top-management's agenda. Broad information and involvement of all employees, to the point of 'communication overkill,' is an integral part of the proceedings.

SETTING OBJECTIVES – IT'S WORTH THE EFFORT

Nothing frustrates improvement programmes more easily than unjustified or unjustifiable objectives. If goals are perceived as unrealistic and unfounded, insuperable obstacles mount up all by themselves. On the other hand, over-modest goals are often seen as evidence of management weakness, and are equally ineffectual.

The goals must be convincing, and they must convince the people who have to meet them. How can employees invent and implement improvements if they do not know the real extent of the damage, and therefore do not believe in the need for change? Equally important, the goals must be concrete and graphic. The resolve 'We're headed for the world class' may be enough to kick off an improvement programme, but it will not take a company to the top. How can you move mountains if you do not know where you are supposed to put them?

Stories about the successes and superior achievements of others and targets set by management are much less valuable than seeing for yourself. Will a 30 per cent cost reduction do the trick, or do all costs need to be halved? Will productivity have to be doubled for the company to approach the world class?

It is worth taking time to understand the competition, to identify world-class cost levels, world-class standards in productivity improvement and growth rates, and discover the rationale behind them. The effort will pay off; seeing with your own eyes and personal experience often administer a salutary shock.

There are many ways to identify world-class standards. The most important ones – reverse engineering, benchmarking analyses and benchmarking trips – have proved their value in many catch-up programmes. However, the view in the competitive rear-view mirror may only be used to draw forward-looking conclusions for in-house reforms and objectives.

Reverse Engineering

Competitive products are often taken apart; unfortunately, however, the usual reason is to provide sales with evidence of the disadvantages of competitors' products. Useful information on costs and manufacturability is often wasted, it is not processed or used to extend the company's own knowledge.

Yet there is no better opportunity to learn at first hand whether equally good or better solutions can be achieved at lower cost. This has been proved by many reverse engineering exercises, from computers and telephones to white goods or autos or diesel engines.

The most frequent finding of the mixed teams of development engineers, manufacturing staff and buyers, marketing and sales people who have analyzed the cost and benefit of individual product functions and subsystems and compared them with their own product is this: simply optimizing product design will often bring down manufacturing costs, even in Europe, by 25 to 30 per cent.

When a company is under pressure to slash costs radically, each new piece of knowledge appears in a different light. Every previously disregarded detail becomes an opportunity to reach world-class cost structures and to get back on the attack in the volume segment with high-performance products. Excessive material and product specifications that cannot be passed on in the price, and technical solutions that require an unusually high level of effort in manufacturing, quality assurance or servicing, push up costs. In many cases, products were found to have too many parts, overlong cables, hard-to-assemble units, time-consuming test procedures, or to be susceptible to breakdowns (see Figure 8.2).

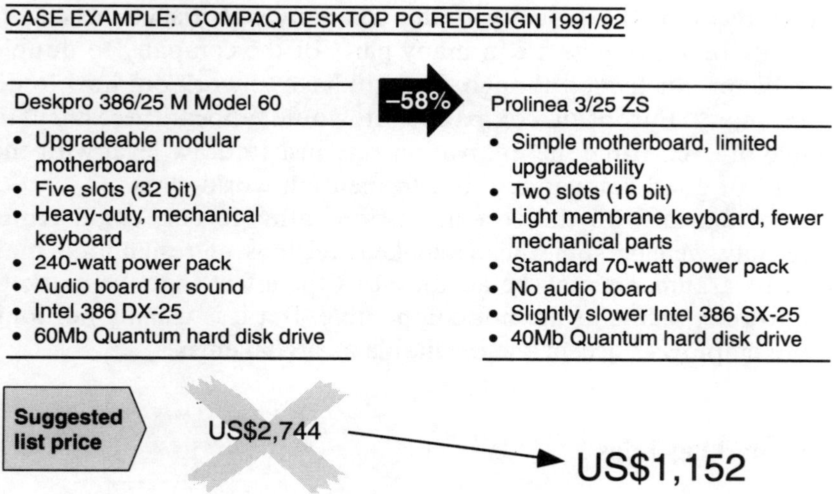

Source: *Business Week*, 2 November 1992

Fig. 8.2 *Design-to-cost provides similar customer value at drastically lower cost*

Benchmarking

This serves to reveal differences in performance structures in the greatest possible detail. It works like a jigsaw puzzle. Data from publications of all kinds have to be collated and analyzed, and interviews conducted with dealers, customers and experts, or even directly with competitors.

Here, too, the company forms mixed teams. Employees should arrive at their findings themselves, to set the right objectives based on what they themselves have perceived. It is important here to counteract discouraging experiences and guard against the premature conclusion that there is no basis for comparison.

There are almost certain to be a few surprises. The 'world-class' companies studied will employ only a third as many production staff to manufacture comparable volume, and the indirect areas will need only a quarter of the people. The analysts will find that many more new products are introduced with fewer development engineers. Successful electronics companies will be achieving 40 to 60 per cent of their sales with products that are less than a year old. They will be growing much faster than the total market. And much more besides.

The raw data sometimes have to be corrected after a double check, but more often the casualty is the company's cherished beliefs. It suddenly becomes possible in many parts of the company to double or treble productivity – though they still have to work out how to do it. In many European companies, it would appear necessary to double or even triple the innovation rate just to draw level with the average of good companies – not to mention world-class.

Here too, as with the cost discussions after reverse engineering, there may well be some heads shaken, feelings of resignation may arise, or arguments may be advanced to the effect that the markets and product segments are not comparable. But it is usually not long before employees accept the inevitable as a challenge.

Benchmarking Trips

Ending comparative surveys with such visits to the world's best industry players will confirm all suspicions. The missing pieces of the jigsaw fall into place, the last doubts about the findings are dispelled. A huge amount of energy can be released if the incredible results of the comparisons are confirmed at the very source and excellence is seen on site to be fact, not fantasy (see Figure 8.3).

Production employees who, before going on a benchmarking trip, declared that quality goals of 99.5 per cent were totally absurd (having just achieved an improvement from 90 to 95 per cent) became convinced after visiting facilities in Singapore, Korea, Taiwan and Japan that 99.5 per cent was the level needed. They had also gathered plenty of ideas about how to achieve it. At the same time, they brought back from their visit ideas for product design improvements that would allow higher process quality in SMT assemblies. And for the first time they understood why one employee is in charge of just one machine in Europe, whereas world-class benchmark companies manage five to seven.

Overall, benchmarking trips have proved a good investment. Business unit teams come back full of enthusiasm every time, urging management to announce and attain a world-class goal originally regarded as impossible. The spontaneous reaction of one top man-

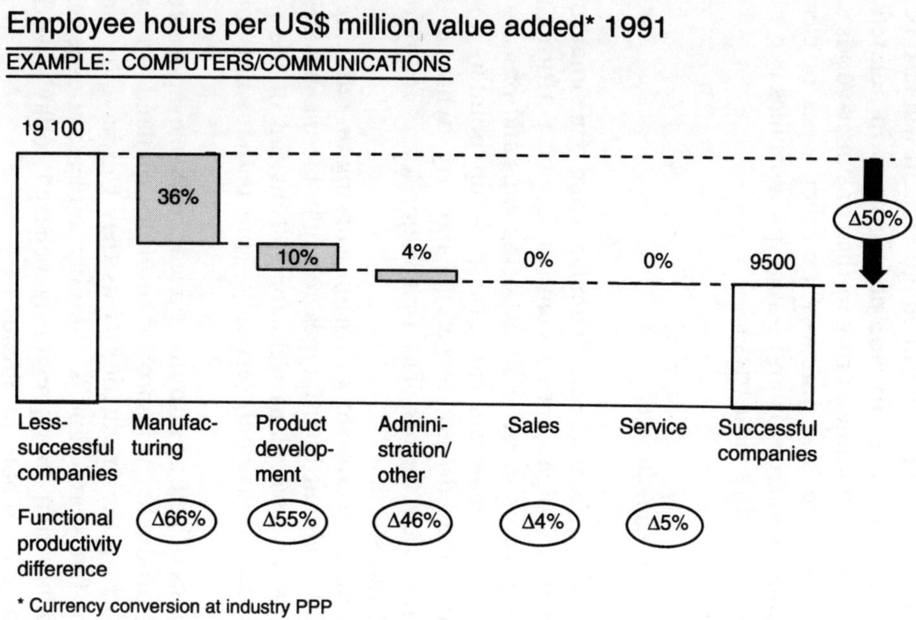

Fig. 8.3 *Productivity gaps of less-successful companies by function*

ager to such a wave of enthusiasm was: 'We only made one mistake, we didn't send enough of our development engineers on these trips.' The next reaction was a second world-class trip, in which 60 per cent of the travellers were development engineers.

Benchmarking visits to competitors with factory visits arranged by both sides or joint workshops on industrial topics are the best variant of such activities. However, a lot of enlightenment will also be gained from benchmarking visits to non-competitors in related areas. Printed-circuit-board manufacturing in the consumer-goods segment gives indications as to best practices for other types of electronics production; quality standards and testing philosophies can be taken over from neighbouring industries.

Forward-Thinking Conclusions

Reverse engineering, benchmarking analyses and benchmarking trips give very concrete insights into a company's own performance and that of its competitors, mercilessly exposing gaps and omissions. But beyond the audit, observers can also pick up pointers to future developments – they do not forget the industry dynamics for the look in the rear-view mirror. The first major levers for reform are developed at this stage.

One visit found a Matsushita division working on reducing its 800 different printed-circuit-board types to eight in one business. The visiting company's own plans were for 300 instead of the previous 3000. A very ambitious goal, certainly – but perhaps not ambitious enough.

Many dreams of getting into the top league have evaporated in the light of the productivity improvement rates chalked up by good companies. Seeing the production costs they will have to slim to in three, four or five years' time is a doubly unpleasant experience for many companies that had previously believed today's world-class cost levels to be rock bottom already.

The convincing and realistic goals needed may easily fall through the cracks between overconfidence and timidity. The best remedy here is to get a full and complete picture of your own core costs. If all defects, delivery times and complaints are eliminated, if everything is running smoothly – running ideally – core costs will remain of 15, 20 or 30 per cent of current costs, depending on the industry and product segment. The conclusion is obvious: even the world-class is far from perfect – even the best players still offer enormous potential for improvement. It also becomes quickly evident that standing still means losing ground; every year that passes without action widens the gap to world-class.

Such costly comparative programmes will have fulfilled their purpose if they convince employees that the company needs a 'general overhaul': that work must start at once on a total redesign, a rewrite of all processes, a review of all structures and product concepts, in order to close the gap to the world-class and not get into a situation where the only option is to close down the business.

The shock triggered by the realization of the gap to the world's best, as revealed by comparison, has to be converted into positive energy. And this is most likely to work if top managers carry out a thorough review of the fundamental parameters and then unite behind a benchmarking result to declare the best-practice models as their goals. The energy gained from the benchmarking phase will normally be dissipated if weeks and months are spent in a large group discussing the comparability of business details and making doubts about details an excuse for protracted delays before starting the total programme.

Detailed objectives are not finalized at the time when the comparisons are completed. This generally occurs when the programme is being drawn up, the transition is not clearly marked.

TOTAL PROGRAMME – TIME FRAME AND ROAD MAP

In successful catch-up schemes, all critical steps are planned in advance and growth and restructuring programmes are developed simultaneously. The change process takes precise, systematic planning. Otherwise, in view of the large number of issues to be managed, it would be easy to lose sight of objectives; there is a risk that resources may be fragmented. Setting goals for restructuring and growth in advance forestalls the possibility that people will start to raise doubts about the seriousness of the situation as soon as restructuring progress and economic revival produce their first wins.

There is no better formula for catching up and overtaking than to map out the route clearly, put the organizational tools in place, and announce restructuring and growth programmes simultaneously. This adds up to five rules for an action programme:

1. Establish Overall Perspective and Manageability of Scope

'Total redesign to world-class' programmes have been successful at major companies with sales of twelve, six or three billion deutschemarks. However, such programmes are more concrete and understandable for all participants if they are carried out in business units with sales of DM500 million to DM1 billion, where the business and the relationships between individual programmes are understood, at least by all second- and third-level managers.

These manageable units are most likely to produce committed teams capable of taking market share from the competition quickly with their reforms. In this situation, all participants are well enough acquainted for the manufacturing man to criticize the development engineer for his poor product design, the problems it causes in assembly, or the exotic mix of component structures on the circuit boards.

In small units everyone will know that 'We have to change everything to reach our target.' And holistic action becomes a motivator. The team places more trust in a broad-based programme than in a tentative single project. Management has more credibility if it demonstrates clearly that it has considered every individual task. Units that get on track for the world class stand as the best models for similar programmes in other parts of the company.

2. Give Objective, Transparent, and Complete Information

The more familiar the team is with the imperatives of the situation and the necessary redesign programmes, the easier it will be for it to complete challenging tasks. This has been proved again and again when seemingly doomed businesses are turned around. Every team will take the route that saves its own life if it knows exactly what the key issues are.

It is therefore important to communicate world-class goals with the same care and attention as went into setting them. The objectives must be well-understood by everyone, preferably announced by the chairman of the board in person and given credibility through fast decisions that act as clear signals. Reluctance to face unpleasant truths has to be boldly overcome. Only then will the objectives act as a guideline for action throughout the company.

3. Who's Afraid of Too Many Task Forces?

'We have customers too', 'We have a job to do as well' – are cries often heard in companies starting total redesign programmes for the first time. If one improvement programme succeeds another, with the clear aim of reaching the top, such complaints soon subside. Boosting productivity *is* then the job to be done.

In any all-out effort to catch up with the world class, the total programme has to make optimization projects a permanent fixture. The line takes responsibility for this, supported by cross-functional task forces that will also become a fixture in fast-changing companies – albeit with changing membership and responsibilities. In many cases, ten or more task forces at a time help to drive improvements and secure a world-class position.

4. Exploit the Love of Technology

The findings of reverse engineering have an extremely stimulating effect because they challenge the professional pride of the development engineer. Who wouldn't make an extra effort in this case, and want to prove himself a match for other competitor in product design? This is why design-to-cost campaigns are often used to start off restructuring programmes.

An intelligent, manufacturable, easily serviced, quality-oriented product design is generally the sole factor needed to close one-third or sometimes as much as one-half of the production cost gap. Putting competitors in their place in this way is the first challenge for the engineer. Naturally, these programmes have a lead time before their full effect is visible to all. Still, it is often possible to celebrate first wins from design-to-cost programmes at an early stage.

The sequence of steps depends on the corporate situation. If the company is a 'concealed' turnaround case, where a profitable company is lagging well behind world-class levels, outstanding engineering work may well provide the first shining example of improvements achieved.

Visible wins have a contagious effect on other programmes. But design-to-cost cannot always form the vanguard. When business units are making losses, that is, in the conventional turnaround scenario, immediate 'first aid' is indicated: efficiency improvement in direct and indirect areas – maybe even including plant closures – must go hand in hand with design-to-cost, product-line streamlining, and innovation.

5. Announce Growth and Restructuring Campaigns Simultaneously

Starting growth and restructuring programmes together on the very first day often seems to be asking too much; it may well be appropriate to stagger them by six months. Sometimes, too, it makes sense not to start a growth programme until concepts for new low-cost, high-performance products are available and R&D has already started work on them.

On the other hand, higher growth targets are an expression of the same intensified aggression that a company shows when it pushes restructuring. There are other reasons, too, why a close relationship must be maintained between growth and restructuring. Without growth programmes, the social impact of restructuring would be too severe: cuts to the bone without long-term prospects cannot motivate. If restructuring and growth are planned in parallel – and not separated by an interval of several years – the company is also better prepared for 'unforeseen' events, such as an unexpectedly steep slump in prices or new product offensives by competitors.

Management is therefore well-advised to prepare a total programme consisting of both components at the start of a new era, and also to present it to the employees. Simultaneous commitment to restructuring and growth enables the company to strike a new tone and demonstrate its sense of responsibility.

The restructuring programme will have the clear objective of closing the static and dynamic gap, expressed in money terms, within a given time. Programmes and projects will be listed, their impact estimated, and a profit improvement figure put on them. Management will explain that the job is not complete until the gap has been completely closed.

The development of the growth programme is similar. Here, the labour utilization gap caused by the restructuring programme needs to be closed as far as possible, that is, the loss in value-added must be compensated for by increased sales.

'Total redesign' programmes of this nature are not all the same by any means, although they are almost always composed of the same elements: the main levers of the restructuring programmes are reorientation of the total organization towards fast change, overhauling of product lines and single products, and the improvement of operations, particularly internal processes. Growth programmes concentrate on increasing market impact with new, low-cost products, improving exploitation of existing customer relations, and expanding into new markets (see for example, Figure 8.4).

RESTRUCTURING – FAST ACCESS TO NEW IMPACT

Restructuring programmes are specially tailored with a large number of company-specific sub-programmes and detailed implementation planning. The sequence and duration of the individual projects is usually determined when planning the total programme.

Successful companies start with the programmes that will give them the greatest success fastest. And they are not afraid to wield the knife where necessary. They select the order of projects so as to cement the sense of a new era with early wins, so that one project builds on another and a second wave of adjustments does not become a foregone conclusion. Each of the restructuring measures helps lay the foundations for meeting growth targets.

Refocusing the Total Organization

In large functional organizations, should the entire company be broken up into small business units at the beginning of a restructuring programme, or is it better to wait? This is a question that often concerns management in real turnaround cases. They fear that reorganization will delay the whole process of restructuring too much, wasting valuable time and making it even harder to catch up.

Their fears are usually unfounded. Structural adjustment does cause a certain amount of delay at first, but the months that pass before the new organization is in place are time well spent. Once small, independent units are in place, to the surprise of many people they soon rise to outstanding form and develop a new dynamism. Small units react faster to obstacles in the restructuring process or sales slumps, and add parts on to the programme more quickly to be able to adhere to improvement goals and time schedules. By their second year, these units have made up for the original time loss.

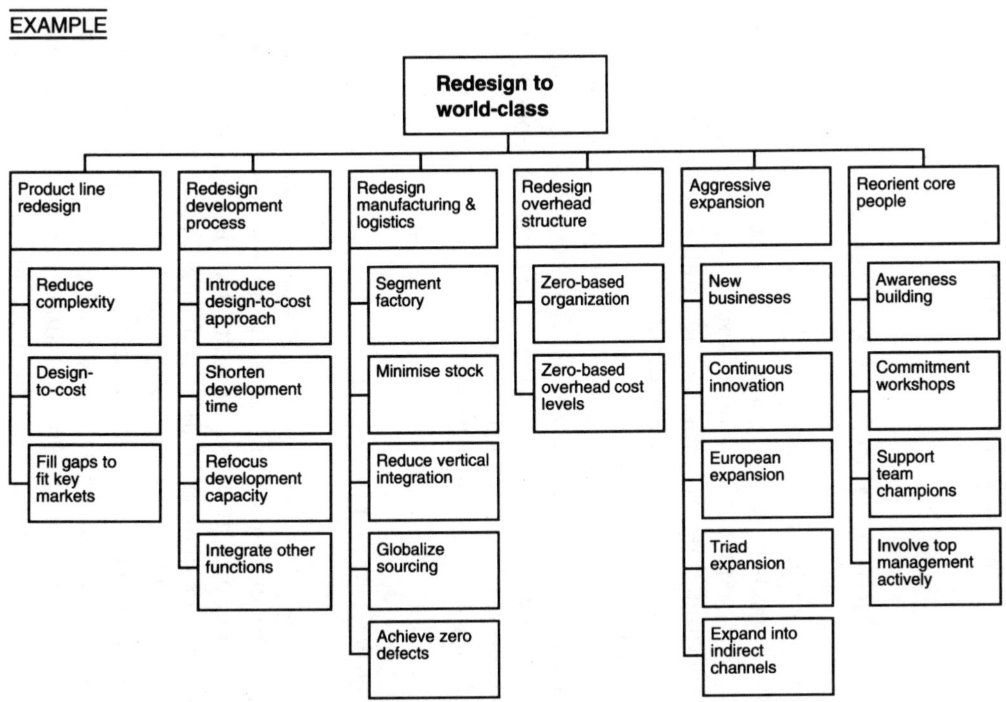

Fig. 8.4 *Typical 'total redesign to world-class' programme*

If a company decides against immediate reorganization it is usually undertaken three or four years later, at the end of the catching-up process. Huge functional organizations then force themselves through the entire improvement process, controlled from the managerial heights in every last detail. The company fails to use what is surely the most effective lever for realizing its world-class ambitions.

Overhaul of Product Lines and Single Products

After organizational changes have been completed, it makes a lot of sense to start the next restructuring steps by creating efficient, low-cost products and pruning the product catalogue. The success of such programmes largely determines what capacities will be needed in development, manufacturing and order-processing and whether expansion projects have any chance of success whatsoever.

The volume market offers the biggest opportunities, and is where there is generally a lack of competitive products. Consequently, product redesign efforts should start at the bottom end of the product range.

'We are gnawing away at the upper segments from underneath,' was how one telecommunications engineer put it. 'We are causing the collapse of the top segments,' enthused the project team. And that really is the case. The idea of developing a 'common family' from 'stripped-down' versions of high-end products is a thing of the past. Once the low-end product has proved the better platform for mid-range business, the top segments will soon be challenged too. Old doubts about whether specially-designed architectures are needed in the middle segment are now strengthened, leading to a radical simplification of the product line.

Once the word gets around that a systems company's mid-range products are available at half the price of the original and are being sold in previously inconceivable quantities, there will soon be imitators in the company's other business units.

Pruning product variety is the next important issue in the electronics industry and far beyond it – in cases with three to ten times more variants than the world's best players, connoisseurs of complexity will not be surprised to see exploding costs and negative returns on sales. Standard products burdened with too-high costs subsidize niche variants. A vicious circle is created: we lose volume in the market, our customers fail to grasp the cost disadvantages of the 'tartan tulip', the road to unprofitable variants leads ever more steeply downhill.

Complexity reduction begins with ABC analyses, which reveal order sizes and sales per product and customer. But the real challenge then lies in filtering out the products that, on the basis of the 'true' direct and indirect costs, make a small or even negative profit contribution.

It is quite common for an analysis of this kind to kill off up to 75 per cent of variants, which are often replaced by mainstream products with the change frequently unnoticed or even welcomed by customers. Standardization of parts and modules helps too. Consistent use of certain parts, even if they are too high-quality or overspecified for some applications, opens the way to cost-competitive hit products. This is a principle often used by Far Eastern electronics suppliers, as you will see by looking inside their products. Unpopulated parts of a circuit board indicate that the 'small' circuit boards have fallen a prey to standardization and that the next product upgrade has already been planned.

Alongside streamlining and standardization, pricing-to-cost is a reliable way of adjusting your course. When their true costs are charged to them, some variants appear in the catalogue at twice the price – the customer is almost always the first to make the necessary cut.

The corporate structure is usually changed in the course of a refocusing programme. The business structure is reviewed, different benchmarks set for different businesses and mainstream businesses are clearly separated from the niche operations, which – often after being aggregated in an independent unit – are given the chance to show whether they can survive without subsidies. Where separation proves impossible, a clear-cut organization by segments is introduced within the business units. This applies to manufacturing as much as to development and order processing.

Improved Operations

Adapting manufacturing structures and capacity to the company's new value-added is usually a particularly painful process, since staff adjustments are more radical here than in many other areas unless massive new growth can be generated with hardware. Adapting manufacturing capacities in electronics companies is not a one-off effort, but an ongoing activity. And this is no wonder when you consider that in future a computer company will require only 25 per cent of its manufacturing capacity for 50 per cent higher volume production, and a consumer-electronics factory can turn out the same number of units in five years' time, after repeated product redesign and process improvements with only a quarter of its present capacity.

The cuts needed are exceptionally deep in almost all total redesign programmes in electronics factories – deeper than in other industries because value-added is, in many cases, shifting more and more away from hardware towards software and service.

There are now major overcapacities in the electronics industry throughout Europe, 30 to 40 per cent in some companies. This occurred partly when local-content preferences were abandoned in a united Europe and when country-specific regulations were dropped, as happened in telecommunications equipment. Added to that were value-added losses, as well as hopes of growth that remained unfulfilled. Even assuming that low-cost, high-performance products at the low end will bring additional growth, it will be difficult to maintain today's European manufacturing capacity.

New long-term, integrated structural concepts for manufacturing are unavoidable. This critical future task must be assumed in the first place mainly by the manufacturing director. Under his leadership, teams will prepare a general concept for the individual production sites, separating mainstream and fringe products and giving them a product-oriented layout, and also presenting ideas for an alternative use of locations.

The 'compression' of cross-functional process chains ('Core Process Redesign') is a further highly-important approach used in restructuring. Excellent performance in core processes, order handling, introduction of new products and so forth, makes a company leaner and faster. Shortening lying times, reducing interfaces, concentrating tasks in one person or one department, avoiding double work on different levels or simply 'doing it right the first time' are the goals of Core Process Redesign.

One of the most important core processes in the electronics industry is undoubtedly the process of product introduction, from development to market launch. Optimization here must focus on more innovation for less development cost. Doubling the innovation rate, as measured by the percentage of sales earned from new products or by the sales of new products per development engineer (or even more ambitious goals depending on the corporate benchmarks), shortening development time for electronic consumer goods from 18 to 6 months, halving development budgets for systems in the top segment – these, or something very like them, are the goals dictated by the market to most electronics companies. They express the need for an efficient new product introduction process.

In successful companies this process is distinguished by cross-functional product teams, good project management and low complexity. Successful innovators prefer step-by-step solutions to complex large projects; they build in quality as early as possible and avoid expensive repairs at a later date; they encourage teamwork between all functions and foster a holistic business perspective – not purely the developer's eye view.

The first step is usually to weed the development project portfolio thoroughly. In this focusing process, variants and improvement programmes for old products are scrapped, and entire product segments dropped. This frees up capacity to make room for product redesign projects.

The task force that aims to create fast, lean product introduction processes will have to cast its net wide. It will probably find scope for improvement in concept development and modification processes, cross-functional product management, reverse engineering to identify the best designs, integration of suppliers into software and hardware development and into system tests, and much more. The measures to speed up the development process are tested in pilot projects and then adopted by all cross-functional development teams.

The development of this action programme is followed by increased concentration on personnel and organizational issues. To ensure that the previously separate functions really are integrated, it is important to have development take over manufacturing tasks or give design work to the factory, for example, the SMD (surface-mounted device) manufacturing team. Marketing and development tasks must be pooled in the large customer business for electronic consumer goods, product line managers must be established and the decision-making powers of cross-functional teams must be extended.

Measuring success is rarely as important as it is here. If the programme is to sustain its momentum, it is vital to see the development time for SMT subassemblies shortening project by project, system tests and system installation taking less time, suppliers reacting faster, concept phases shrinking and the degree of maturity at individual milestones increasing from project to project.

GROWTH – AGGRESSIVELY ON TRACK FOR EXPANSION

Special programmes will be needed to provide an additional innovative growth stimulus. It is worthwhile here to start crash programmes where, perhaps, ideas long-discussed can finally be put into practice, and to separate a star team recruited from all functions from the rest of development in order to close a product gap or save a threatened segment in the least possible space of time. A hundred million marks of additional sales on first entry into a new product segment and a hundred million more for follow-up products are not impossible; a sum like that will pep up the complex restructuring and growth programme.

But even before the development results are available, the first market- and sales-oriented programmes to promote growth can be started.

Specific Sales Concepts for New Products and Markets

Special marketing campaigns for the new 'low-cost/high-performance' products make effective kick-off programmes. Entry into the high-volume business means that the sales department has to adjust to new customer groups. New distribution partners are sought and favourite old strategies are abandoned. Cooperations may also make sense in this situation; shared sales functions and product exchanges should be considered.

Low-priced products open up access to previously closed national markets. Successful companies develop special entry strategies for these countries and draw up business plans. They check whether it will be worthwhile to establish an independent distribution function, or whether it would be better to enter into a partnership.

Product- and market-related considerations pass over into a general discussion of the regional distribution structures as well – in many companies these have scarcely changed over the past 15 years. Direct sales in particular will benefit from a fresh evaluation of today's largely regional structures, and from a complete redesign in some businesses. In other cases simple reorganization may be enough; for instance, six regions instead of eleven.

Reorientation of the Acquisition Process

Extremely regionalized, deeply-hierarchical sales units, few indirect channels, little use of mass distribution channels – many companies ignore the major potential in these areas.

It is best to begin here with a systematic segmentation of the product range. Dividing the assortment into products that need more explanation and products that need less, as well as into standard products and system sales, mass business and project business, can be used as a basis for selecting the most efficient sales method for each business. More use of retailers and mass-distribution channels for the product business, telemarketing for retailers, mail-order and so on, lighten the burden on the sales budget and open up new customer groups. Distribution costs fall from 25 to 10 per cent of total costs.

A combination of imagination and a cool, calculating brain is needed. Tomorrow's distribution concept could look like this:

- Indirect channels will always require central solutions;
- Small retailers and medium-sized service houses will accordingly be served by a central, in-house distributor;
- Direct or system sales will be structured by industry instead of by region, and have their headquarters in each case in the area with the highest customer concentration;
- National sales organizations will frequently turn into European sales networks.

The task forces reorganizing the acquisition process will have more in mind than cost reduction. They will also be planning the return of sales to its true function – away from administration, towards the customer. The fact that sales spends only 30 per cent, or even less than 10 per cent, of its total time at the customer is a clear alarm signal, as is having five sales staff at the office backing each worker in the field. Here, tasks need to be redistributed between customer service people and the back office to strengthen on-site presence with small, integrated teams from office sales and field sales. The creation of special teams to look after key accounts and the use of profit contribution-based incentives are other effective means of increasing productivity in the sales department.

Sales Stimulation Programme

Successful companies see strategically-planned and executed sales stimulation programmes as an important part of every growth programme. Such utterances as: 'No one can teach us how to sell, sales does what it can; no one can do it better' – are not good enough for these companies, and they carry out targeted projects to prove the contrary. Customer by customer, prospective customer by prospective customer – all are closely scrutinized. Detailed information is collected on customer requirements and customer value, possible services, the position of competitors, as well as on the customers' requirements and ordering behaviour in the years to come. This database is used to formulate goals and develop ideas about how these goals can be achieved. Requirements, including investment, have to be discussed. It is not impossible to improve customer value and double sales with existing products when this method is used.

A well-structured package of projects must be introduced for new product distribution, reorientation of the acquisition process and the sales-stimulation programme. Depending on the task in hand and the new organizational structure, they are the responsibility of the sales director, the individual business units or the regional divisions.

THE RUNNING OF THE PROJECT – GIVE VISIBLE SUPPORT

Motivating task forces composed of committed personnel is one thing. But when 2000 or maybe 20 000 employees, some of whom are afraid of losing their jobs, are to be persuaded to take a new direction when they personally could be losers, the problem gains a whole new dimension.

What do you have to do to ensure a smooth change process with no major hitches? Success or failure depends on the commitment of the top-management team. There is no hope of successfully completing a major catch-up programme if the top managers fail to lend their support when employees are being given information and suitable personnel are being selected for project work, or if they are not constantly there to renew the momentum before enthusiasm flags.

Information to the managers starts off every such programme. Managers have to hear at first hand what is planned, what is expected of them, and what they can hope to gain. Failure to provide information generally costs companies dear. Rumours go around, insecurity and anxiety grow that may paralyze the whole organization. The opportunity to win over employees for the cause is lost.

The CEO gives the starting signal. He informs all line managers and project leaders of the results of the company diagnostic, and presents the world-class standard and the targets for closing the gap. This first management meeting is followed by a series of workshops and written notification to employees in each business unit or department.

As the work proceeds, a whole succession of meetings gradually penetrates through the company, with information sessions large and small for employees at all levels. Line managers and project leaders inform their departments after they have further detailed the programme, defined goals for the individual projects, and identified the main points of attack. Kick-off meetings for individual projects follow with the relevant line managers. Soon after, the programme is under way – ten or so projects are running and a large number of teams have started work.

But this is by no means the end of the information process. There are progress meetings at regular intervals during which line managers present projects, the actions developed and their effects, and report on how far they have progressed towards their goal. All project activities and achievements are openly revealed, including all errors.

The top management is scrupulously observant here, looking for true achievement and not impressed solely by good presentations. Stagnation or attitudes of rejection will not remain hidden from them, and they will firmly insist on improvements where they see deviations from track. They will ensure that the competition between the various teams is clearly visible. Lining up ten units or task forces alongside each other in reviews has speeded up many changes.

Cross-functional teams are formed for the individual projects, supporting the line on detail work. The unbreakable rule in staffing these teams is only to appoint people who can think ahead, and only to call on the best – including the best young talent. There is no room for doubters and foot-draggers.

The process is under way, the company is all set for the jump to a new era, and the initial motivation has been generated. Employees have realized that the process will take two to four years and that there can be no going back to the old style. But the fact that they have heard the message does not mean they have internalized it. And still less, that they will act accordingly.

How can motivation be sustained in the long term? A refresher course will be needed from time to time, setting the tone for life in the world-class and permanent self-renewal cannot be repeated often enough. But top management can do more.

Many companies put their faith in broad-based mobilization. General Electric held 'work-outs' with management and a selection of employees down to the bottom of the hierarchy. This involved several thousand employees in the mobilization process. In small, manageable, carefully composed groups, they discussed shortcomings and improvement opportunities, and made binding agreements on these.

Another effective mobilization tool is switching top management positions. Even a good works manager will find it difficult to challenge the structures he has built up over 10 years, literally destroy them and then put them up all over again. That same works manager will have no such inhibitions in a colleague's area. Here, he can go to work confident that he will not be called on to justify past mistakes. Sideways moves in management are often a much more effective way of starting afresh than the carrot and stick method.

Although the remedy should only be used very sparingly, it would be wrong to close your eyes to the facts: a top-management decision to replace the entire management team in one business unit and move the managers to other parts of the company accomplished more change in a matter of months than all previous efforts had done in years. And many R&D departments have also benefited from a change in leadership. Transferring an aggressive innovator to the development department of a neighbouring business unit, where four or five good people went with him to key positions, made a hide-bound area adopt a completely new working approach – in project management, in its attitude to low-cost products, and in its collaboration with service and manufacturing on design-to-cost.

This, give or take variations, is how some companies set about closing the world-class gap. During the process, nearly all found that the entire restructuring and growth programme only appeared to overtax them at the start. In the first year there were complaints, but in the second they grew fainter. Instead, ideas and practical suggestions were offered for major steps forward in areas not yet explored. By that time, task forces and improvement programmes were part of everyday corporate life.

And the question 'Will it never end?' is no longer asked. No, it will never end, and more and more people are coming round to this point of view. The step towards continuous renewal has finally been taken.

Appendix A: World-Class or Not? A Manual for Self-Diagnosis

The long-term study 'Excellence in Electronics' reveals a 'one-third society' forming globally in the electronics industry. Only the top one-third of companies has good future prospects in the medium term; medium-rank and less-successful companies must change course fast and radically or else drop out.

The only protection against a company overestimating its own corporate strength is an objective diagnostic that will mercilessly expose any weaknesses. Is the company reaching the productivity and growth levels of world-class competitors, or of comparable companies in other sectors? Are the returns to date high and not just 'acceptable'? Are high returns due to genuine competitive advantage or to the fact that the markets are – for the moment – still protected? Is the company exporting to the big Asian growth markets, or is its outstanding position abroad concentrated in a handful of neighbouring European countries? How much does the company need to improve to be world class in a few years' time? What does this improvement goal mean for all areas of responsibility, including the lowest? What levers have to be applied to close the gap to world-class successfully?

Initial answers to these questions and indicators of the main (weak) points that need to be examined in greater depth are revealed by the long-term study:

- The study's findings were used to draw up a *checklist of questions* which look at the main differences between successful and less-successful electronics manufacturers.
- A number of key, easy-to-apply *metrics* are available for evaluating the answers.

THE CHECKLIST: KEY FACTORS FOR SUCCESS (KFS)

The ratios needed to assess a company's situation can be determined using the checklist following in Appendix B. It pulls out not only data from the individual corporate functions – such as R&D, manufacturing and purchasing – but also overall business data and cross-functional behaviour patterns.

The checklist – also applicable outside the electronics industry, with slight modifications – is divided into six areas:

- *Overall business situation* (return on sales, sales growth, world market share, productivity and so forth);
- *Focus and complexity* (number of variants, number of customers, suppliers, development projects);
- *Manufacturability and design-to-cost* (number of components to be inserted, assembly steps);
- *Operational efficiency of product development* (innovation productivity, R&D expenditure);
- *Operational efficiency of manufacturing* (insertion productivity, capacity utilization, quality);
- *Characteristics of a problem-solving organization* (competitive comparisons, cross-functional teams)

The questions are based on the key factors for success in the electronics industry as revealed in the long-term study. These are the parameters in which statistically significant differences were found between successful and less-successful companies as shown in Table A–1.

It should be remembered that the long-term study focused on manufacturing and development. The checklist therefore does not directly address the areas of sales, marketing and customer service. Likewise, since the study relates to individual business units or single-product companies and their existing product ranges, the checklist could not feature the setup of completely new businesses either. These issues will be examined more closely in a second stage of 'Excellence in Electronics'.

METRICS: THE FINDINGS OF THE LONG-TERM STUDY

Some of the answers from the self-diagnosis can be directly compared with corresponding data from the long-term study (compare the numeric values in Table A–2 and in the checklist). These data can help give companies initial indications of the areas where there is most need for improvement. Further key data such as the levels of labour and development-productivity can be taken from references in the text or individual Figures. In such cases, however, the results should be checked from case to case, since there may be differences in the way data are defined and calculated.

Table A–1 Parameters which were more marked at leading companies to a statistically significant degree (KFS)[1]

Significance	*KFS*
(A) Overall business situation	
< 1%	Higher world market share
	Higher labour productivity
	Lower share of manufacturing cost
	Lower workforce cuts
1–5%	Higher percentage of sales with superior products
	Fewer indirect manufacturing employees per unit of value-added
	Fewer development engineers per unit of value added
	Fewer sales personnel per unit of value-added
	Higher labour productivity increase
5–15%	Positioning stronger in growth areas
	Higher share of sales in recently launched products
	Higher unit-cost reduction
(B) Focus/complexity	
< 1%	Fewer module variants per unit of sales
	More inserted components per PCB assembly line and year
1–5%	Less customized manufacture
	Fewer development projects per unit of sales
	Higher purchased volumes per supplier
5–15%	Fewer product and part variants per unit of sales
(C) Manufacturability and design to cost[2]	
1–5%	Higher percentage of SMD use in PCB assembly
(D) Operational ratios in product development	
< 1%	More sales with new products per development engineer
1–5%	Lower ratio of R&D spend/Sales
(E) Operational ratios in manufacturing	
< 1%	Higher process quality in final assembly
1–5%	Higher PCB assembly process productivity
	Higher capacity utilization
5–15%	Higher process quality in PCB assembly
	Shorter manufacturing throughput times

Table A–1 continued

Significance	*KFS*
(F) Problem-solving organization	
< 1%	Higher development spend per development engineer
1–5%	More unit-cost reduction targets based on external sources Greater number of development engineers in the team full-time Suppliers with better performance More active and long-term partnerships with suppliers Tighter supplier linkage to the manufacturing process Greater emphasis on cooperative approaches for quality improvement and reduction in purchase prices More formal project milestones More structured procedures for software development Higher frequency of genuine development teams with strong project managers More machine operators with direct problem-solving responsibility
5–15%	More active use of external sources to give development engineers ideas Lower automation expenditure More incentive systems even on the lowest level

[1] Classified by return on sales and sales growth
[2] Direct product comparability only possible in small samples, so little or no significant data here

Detailed findings and ideas for concrete improvements should be taken from specific analyses, such as benchmarking visits to world-class competitors or systematic reverse engineering of the best competing products.

When interpreting the comparative data shown here, the following points must be considered (as they were in the long-term study):

- The *differences between the individual segments* of the electronics industry;
- The high *dynamics of change* in the industry;
- International *comparability in currency conversion.*

Differences between Industry Segments

All ratios are given for the four segments: industrial electronics/measuring systems, large systems, computers/communications and consumer electronics/small products. Despite the wide variety of types of products in each segment – the computers/communications segment, for example, combines such varied products as PCs, minicomputers, printers and private automated branch exchanges (PABX) – the most important ratios show amazingly homogeneous features. In individual cases, as with data on development times, there is a sharp divergence; analysis at the product group level is needed here.

The individual company should first compare itself against the ratios of the industry segment most similar to its own, in order to draw conclusions about its position versus competitors.

In addition, a comparison should be made with the next most mature industry segment, as there is much evidence to suggest that this may be the next yardstick for successful companies (see Figure A–1). This means that in the computers/communications segment, for example, successful companies have to orient themselves to the segment with the greatest degree of maturity; that is, consumer electronics/small products, where very lean cost structures are needed even outside manufacturing in the form of more indirect sales channels, for example. This is the only way to survive the 'shake-out' caused by intensive price competition. By contrast, for less-successful companies whose primary aim is to break into the top group, the immediate goal is to optimize manufacturing processes and manufacturing costs. But these companies also need to be aware of future needs, so as not to be lulled prematurely into a false sense of security.

Awareness of Industry Dynamics

As a rule, the comparative data for successful and less-successful companies are average figures. When using averages it should always be remembered that world-class companies normally lie well above these levels. It is for this reason that, in some cases, the figures for the world-class players are used.

The calculations are based on 1991 figures. The figures for 1988 and 1989 were also collected in some cases, so that improvement rates could be calculated. Projections up to 1994 were also requested, to allow forecasts of future developments. In general, top companies in 1991 had a lead of about three years over the average successful company.

For the yardstick to be right, the results of the study must be updated for the current year. The improvement rates attainable indicate how important this updating is. For example, improvement champions can be identified by the fact that they combine their particularly high restructuring speed with fast growth rates as shown in Figure A–2.

Table A–2 Summary of the most important individual parameters[1]

	Industrial/ measurement		*Large systems*		*Computers/ communications*		*Consumer/small products*	
	Successful	*Less successful*	*Successful*	*Less successful*	*Successful*	*Less successful*	*Successful*	*Less successful*
(A) Overall business situation								
1. Return on sales (%)	19	−7	16	−2	12	−9	5	−4
2. Increase in return on sales (% pts, 2 yrs)	0	−1	2	0	1	−3	1	−1
3. Sales growth 1989–91 (% p.a.)	11	10	12	6	26	−8	6	−4
4. Share of sales of superior products (%)	53	22	48	39	43	35	39	41
	World class: 100		*World class: 68*		*World class: 80*		*World class: 80*	
5. Share of sales of recently launched products (%) (introduced in the last 12 months)	9	11	26	28	49	27	45	28
	World class: 28		*World class: 50*		*World class: 82*		*World class: 80*	
6. Personnel costs in manufacturing (direct and indirect)	21	28	22	43	13	23	11	28
7. Labour productivity increase[2] (% p.a.)	8	5	14	12	13	14	8	−3
8. Unit cost reduction 1991–94 (% p.a.)	12	6	11	10	14	17	8	7
(B) Focus/complexity								
Number of product variants per US$ 100 million sales								
9. Products	402	1337	18	380	46	76	141	176
10. Modules	479	1520	81	427	85	330	133	245
11. Development projects per US$100 million sales	29	62	11	27	7	14	16	25

Table A–2 continued

(C) Design to cost and manufacturability								
12. Share of surface mounted device (SMD) insertion 1994 (%)[3]	70	37	78	29	80	81	41[3]	52
(D) Operational ratios in product development								
13. New products per 100 development engineers 1989–91 (incl. modifications)	22	7	2	21	35	9	23	19
14. R&D expenditure (% of sales)	10	12	10	10	7	16	6	6
15. Development time (months)	Controls 23	Controls 24			PCs 11	PCs 11	TV/VCR 15	TV/VCR 18
	World class: 15				*World class: 4*		*World class: 10*	
16. Time lag between design freeze and market launch (months)			Minis 3	Minis 4	PCs 2	PCs 5	TV/VCR 2	TV/VCR 4
(E) Operational ratios in manufacturing								
Process quality, final assembly								
17. After power-on test (% scrap)	8	26	N/A	N/A	6	20	2	2
18. After function test (% scrap)	4	23	N/A	N/A	3	13	2	5
19. Total output (% accepted parts)	92	72	N/A	N/A	92	76	97	93
20. Throughput time final assembly (hours)	Controls 20	Controls 54			PCs 12	PCs 14	TV/VCR 5	TV/VCR 21

[1] All currency in US$, converted at purchasing power parities, base year 1991
[2] Related to value-added per hour worked
[3] Successful consumer electronics figure of 41. Product design for TVs often dependent on existing production lines (PTH technology)

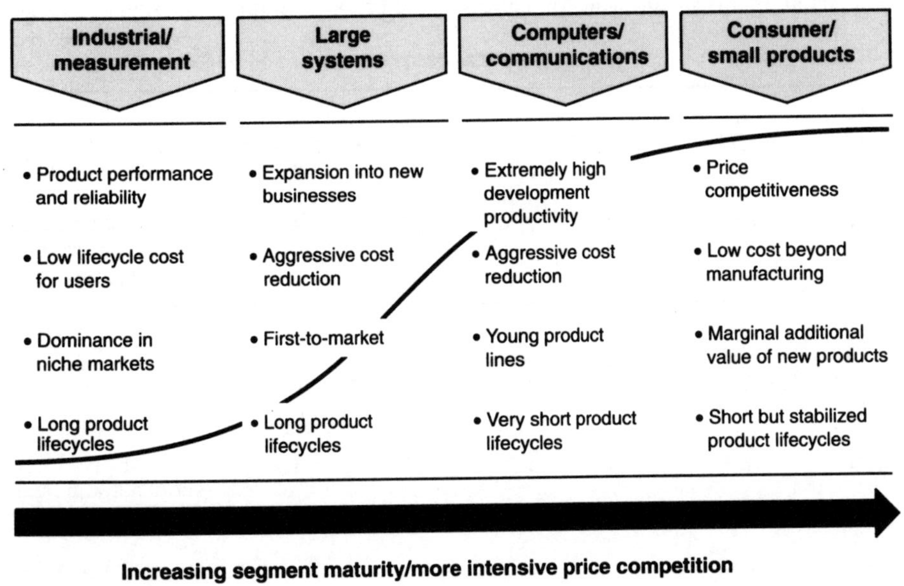

Fig. A–1 *Industry segment-specific characteristics*

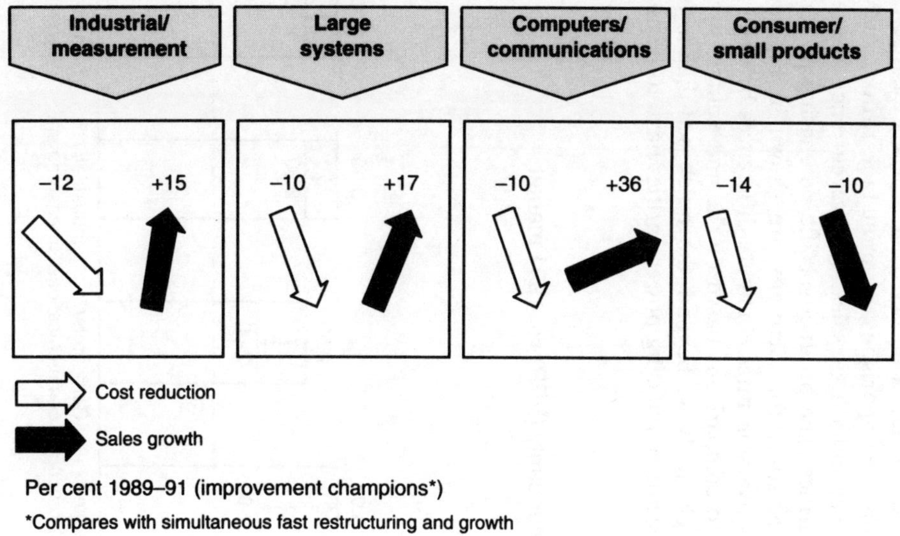

Fig. A–2 *Segment-specific improvement rates for restructuring and growth*

Currency Conversion

The extent to which market exchange rates can falsify comparisons can be clearly demonstrated by the example of cordless telephones for private households. At the beginning of the 1990s, phones were available in the US for a store price of about US$100; in Germany a comparable unit still cost about DM500. In this case, the purchasing power parity would be 5 Deutschmarks to the US dollar, a far cry from official exchange rates.

Correct international comparisons of monetary ratios, such as the number of variants per million of sales, require currency conversion at purchasing power parities. The product-specific purchasing power parities used in the long-term study for Germany are shown in Figure A–3 as bandwidths. These conversion rates eliminate differences between national price levels, leading to comparisons based on the unit volume parameters. For a more thorough analysis, beyond the quick self-diagnosis of the checklist, product-specific purchasing power parities have to be established individually.

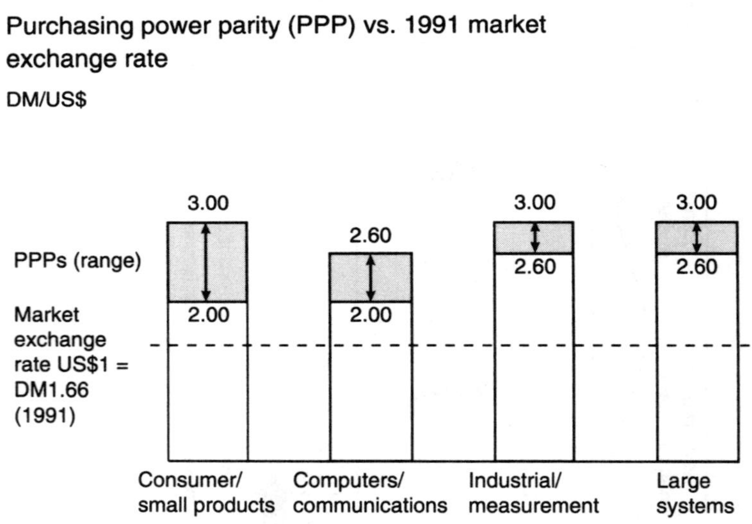

Source: McKinsey analysis

Fig. A–3 *High price levels in Germany*

Measured against the key success factors, is your company world-class or not? The self-diagnosis will give initial indications.

Appendix B: Self-Diagnosis: A Checklist for Benchmarking Against World-Class Companies in the Electronics Industry

(*Figures in brackets refer to comparative data in Table A–2*).

(A) WHERE DOES THE BUSINESS STAND AS A WHOLE?

Is it profitable enough, and are changes upwards?
Return on sales (1)
Increase in return on sales (2)
Sales growth (3)
Is the business operating in growth areas?
Does the main product group have a leading position in the world market?
World market share
Are the products competitive?
Share of sales of superior products (4)
Share of sales of recently launched products (5)
Is current overall productivity excellent?
Labour productivity (value added per hour worked)
In which functions does this not apply?

- Manufacturing
- Direct labour
- Indirect labour
- Development
- Sales

Is the share of processing costs in manufacturing too high?
Personnel costs in manufacturing in % of sales (6)
Are the overall productivity rates increasing enough?
Labour productivity increase (7)
How far do productivity increases depend on workforce reductions?
Rate of change in employee numbers

Or do increases in productivity depend more on increasing the volume of value-added?
Growth in volume of value added
What productivity goal has to be met within three years, allowing for competitive dynamics?
Extrapolation of labour productivity of the world-best competitors
Are unit cost improvements sufficient to achieve this?
Unit cost reduction per annum achieved to date (8)

(B) IS POOR FOCUS LEADING TO HIGH COMPLEXITY?

Does manufacturing have too many product variants to cope with?
Number of variants per unit of sales

- Products (9)
- Modules (10)
- Parts

Is the range of products and parts being reduced fast enough?
Are appropriate methods being consistently used?
Use of component preference lists
Share of active part types on component preference list
Avoidance of customer-specific manufacture
Are too many individual customers being served?
Number of customers per unit of sales
Are there too many suppliers?
Number of suppliers per unit of purchasing volume
Are the production lines focused?
Inserted components per PCB assembly line and year
Are too many development projects being undertaken at once?
Number of development projects per unit of sales (11)

(C) IS DEVELOPMENT GEARED TO COST-OPTIMIZED AND MANUFACTURABLE PRODUCTS?

Are products developed in such a way that as few steps and as little time per step as possible are required for final assembly?
Are components requiring manual work avoided?
Share of manual work in PCB assembly
Is the use of new technologies (SMD) being pushed?
Share of SMD use in PCB assembly (12)
Is the share of costs of (electro)mechanical components minimized?
Manufacturing cost share of non-electronic product parts

(D) DO OPERATIONAL RATIOS IN PRODUCT DEVELOPMENT REACH WORLD-CLASS LEVELS?

Is innovation productivity excellent?
Sales of recently launched products per development engineer
Number of new products per 100 development engineers (13)
Is too much being spent on development?
R&D expenditure in % of sales (14)
Do development projects take too long?
Development time (15)

(E) DO OPERATIONAL RATIOS IN PRODUCTION REACH WORLD-CLASS LEVELS?

Can PCB assembly keep pace?
Insertion productivity: number of components per hour worked
Do capacities meet requirements?
Capacity utilization
Number of shifts per day
Is quality manufactured in or tested in?
Process quality in PCB assembly

- After in-circuit test (first-pass yield)
- After function test
- After reworking
- Total yield

Process quality of final assembly

- After power-on test (first-pass yield) (17)
- After function test (18)
- Total yield (19)

Is inventory turn fast enough?
Raw material
Semi-finished products
Finished products
Are the manufacturing throughput times at world best levels?
Manufacturing throughput time per unit

- PCB assembly
- Final assembly (20)

Is delivery reliability at the high level normal in the industry?
Share of on-time deliveries

(F) IS THE COMPANY A TRULY PROBLEM-SOLVING ORGANIZATION?

Are internal and external parties sufficiently involved?
Are comparisons made regularly with the best competitors?
Benchmarking
Reverse engineering
Do development engineers actively use external sources to stimulate new ideas?
Direct discussions with customers
Sales employees
Service employees
Cooperation with suppliers
Are the goals for unit cost reduction based on external sources?
Competitors' unit costs
Customer preferences
Expert assessment based on external information
Do development engineers make use of results from upstream development/research, and is this done in close collaboration with the upstream developers/researchers?
Are other functional areas largely integrated into development teams full-time and not just part-time?
Purchasing
Production
Sales/marketing
Customer service
Accounting
Are development engineers also committed full-time as far as possible?
Share of full-time team members among the development engineers on the team
Are structured methods for information processing used in the development process?
Use of value analysis
Cost reduction achieved from value analysis
Are these methods applied by the development teams themselves?
Are joint ventures used for product development?
Number of joint ventures in development per unit of sales
Is the development process sufficiently oriented toward problem-solving and responsiveness?
Are targets for development engineers primarily market-oriented?
Are specification changes allowed before product launch?
Number of specification changes before product launch
Are specifications finalized late enough before product launch to stay close to the market?
Time lag between design freeze and product launch (16)

Are expensive specification changes after product launch avoided?
Are there sufficient measuring points in the development process?
Number of formal project milestones
Are there structured procedures for software development?
Prototyping in the specification phase
Explicit trade-off between performance features and costs
Software quality testing as explicit task
Is there enough delegation of tasks and responsibilities down the organization?
Are there genuine development teams with strong project managers?
Are there project teams that are answerable to project managers with disciplinary responsibility?
Do project managers report direct to the business unit head?
Do project managers or development teams have sufficient final decision-making powers?
Examples:

- project budget
- project organization
- design/performance features

Are operational and scheduling tasks and responsibilities in manufacturing largely assigned to machine operators?
Examples:

- quality data gathering
- operational problem-solving
- material ordering

Is delegation of responsibility sufficiently supported?
Examples:

- employee training for quality-assurance assignments
- concrete description of goals and goal-attainment levels

Or is too much invested in automation?
Automation expenditure in % of value-added
Are supplier capabilities sufficiently exploited?
Do suppliers perform well?
Average delivery time
Average delivery reliability
Are low purchasing prices more important than the total optimization of the value chain?
Purchase price reduction in the last year

- electronic parts
- mechanical parts

Cost reduction from redesigns in collaboration with suppliers

Are enough active, long-term partnerships with suppliers fostered?
Share of active partnerships in purchased volume
Share of long-term supplier relationships in purchased volume
Do close supplier links to the manufacturing process function well?
Share of just-in-time shipments in purchased volume
Share of outsourced parts with no incoming goods inspection
Are cooperative approaches to reducing purchase prices used?
Price reductions over time agreed in advance
Joint redesign efforts with suppliers to reduce unit costs
Or are only one-sided measures used?
Price competition between suppliers
Consolidation of the supplier base
Combination of orders to achieve bulk discounts
Are there any cooperations with outside contractors to enhance quality?
Early involvement of suppliers in product development
Regular joint meetings on quality-assurance issues
Are employees sufficiently empowered to make a productive contribution to a problem-solving organization?
Is job rotation for development engineers encouraged?
Share of development engineers rotating jobs in the last three years
Average stay in other functional areas
Does job rotation take place mainly between manufacturing and development?
Are incentive systems used on all hierarchical levels – even the lowest?
Is team performance assessed and not just individual performance?

Index

Note: page numbers in *italics* indicate figures or tables.